Springer Series in
SOLID-STATE SCIENCES 139

Springer
Berlin
Heidelberg
New York
Hong Kong
London
Milan
Paris
Tokyo

Springer Series in
SOLID-STATE SCIENCES

The Springer Series in Solid-State Sciences consists of fundamental scientific books prepared by leading researchers in the field. They strive to communicate, in a systematic and comprehensive way, the basic principles as well as new developments in theoretical and experimental solid-state physics.

126 **Physical Properties of Quasicrystals**
Editor: Z.M. Stadnik

127 **Positron Annihilation in Semiconductors**
Defect Studies. By R. Krause-Rehberg and H.S. Leipner

128 **Magneto-Optics**
Editors: S. Sugano and N. Kojima

129 **Computational Materials Science**
From Ab Initio to Monte Carlo Methods
By K. Ohno, K. Esfarjani, and Y. Kawazoe

130 **Contact, Adhesion and Rupture of Elastic Solids**
By D. Maugis

131 **Field Theories for Low-Dimensional Condensed Matter Systems**
Spin Systems and Strongly Correlated Electrons
By G. Morandi, P. Sodano, A. Tagliacozzo, and V. Tognetti

132 **Vortices in Unconventional Superconductors and Superfluids**
Editors: R.P. Huebener, N. Schopohl, and G.E. Volovik

133 **The Quantum Hall Effect**
By D. Yoshioka

134 **Magnetism in the Solid State**
By P. Mohn

135 **Electrodynamics of Magnetoactive Media**
By I. Vagner, B.I. Lembrikov, and P. Wyder

136 **Nanoscale Phase Separation and Colossal Magnetoresistance**
The Physics of Manganites and Related Compounds. By E. Dagotto

137 **Quantum Transport in Submicron Devices**
A Theoretical Introduction. By W. Magnus and W. Schoenmaker

138 **Phase Separation in Soft Matter Physics**
Micellar Solutions, Microemulsions, Critical Phenomena
By P.K. Khabibullaev and A.A. Saidov

139 **Optical Response of Nanostructures**
Microscopic Nonlocal Theory By K. Cho

Series homepage – http://www.springer.de/phys/books/sss/

Volumes 1–125 are listed at the end of the book.

Kikuo Cho

Optical Response of Nanostructures

Microscopic Nonlocal Theory

With 56 Figures

Springer

Professor Dr. Kikuo Cho
Graduate School of Engineering Sciences
Osaka University
Machikaneyama-cho 1-3
Toyonaka 560-8531
Japan

Series Editors:

Professor Dr., Dres. h. c. Manuel Cardona
Professor Dr., Dres. h. c. Peter Fulde*
Professor Dr., Dres. h. c. Klaus von Klitzing
Professor Dr., Dres. h. c. Hans-Joachim Queisser
Max-Planck-Institut für Festkörperforschung, Heisenbergstrasse 1, D-70569 Stuttgart, Germany
* Max-Planck-Institut für Physik komplexer Systeme, Nöthnitzer Strasse 38
D-01187 Dresden, Germany

Professor Dr. Roberto Merlin
Department of Physics, 5000 East University, University of Michigan
Ann Arbor, MI 48109-1120, USA

Professor Dr. Horst Störmer
Dept. Phys. and Dept. Appl. Physics, Columbia University, New York, NY 10027 and
Bell Labs., Lucent Technologies, Murray Hill, NJ 07974, USA

ISSN 0171-1873

ISBN 3-540-00399-1 Springer-Verlag Berlin Heidelberg New York

Library of Congress Cataloging-in-Publication Data: Cho, Kikuo, 1940– . Optical response of nanostructures: microscopic nonlocal theory / Kikuo Cho. p. cm. – (Springer series in solid-state sciences ; 139) Includes bibliographical references and index. ISBN 3-540-00399-1 (alk. paper) 1. Nanostructures–Optical properties. I. Title. II. Series. QC176.8.N35C46 2003 530.4'12–dc21 2002044657

Springer-Verlag Berlin Heidelberg New York
a member of BertelsmannSpringer Science+Business Media GmbH

http://www.springer.de

Typesetting: Data conversion by Letex, Leipzig
Cover concept: eStudio Calamar Steinen
Cover production: *design & production* GmbH, Heidelberg

Printed on acid-free paper SPIN: 10469256 57/3141/ba - 5 4 3 2 1 0

Preface

This book deals with a recently developed theoretical method for calculating the optical response of nanoscale or mesoscopic matter. There has been much interest in this type of matter system because it brings out a new feature of solid state physics, viz., the central importance of the quantum mechanical coherence of matter in its transport and optical properties, in contrast to bulk systems.

The author has been interested in the optical properies of mesoscopic matter since the mid-1980s, seeking to construct a new theoretical framework beyond the traditional macroscopic optical response theory. The new element to be included is the microscopic spatial structure of the response field and induced polarization, and the nonlocal relationship between them. This is the counterpart of the size quantization of confined electrons or excitons reflecting the sample size and shape in detail. Although the latter aspect has been widely discussed, the former has not received due attention, and this has prompted the author to introduce a new theoretical framework. This book describes such a theory, as developed by the author's present group. Although it is only one of several such frameworks, we believe that it is constructed in a sufficiently general manner to apply to the study of the linear and nonlinear optical responses of nanostructures of various sizes and shapes, subjects of considerable interest today.

This material has been given as graduate lecture course in Osaka university and several other universities in Japan, or as talks in international conferences during the last decade, with additions and revisions from time to time. Although it has long been scheduled for book form, the project has been hindered by a series of unexpected events, including the Kobe earthquake and involvement in university politics. However, the delay has not only brought negative effects. In particular, we have been able to include more examples of applications, demonstrating the usefulness of this framework in various situations. As an example, the recent observation of the degenerate four-wave mixing signal for weakly confined excitons in GaAs slabs, which is multiply resonant with sample size and light frequency, has provided definite support for the predictions of the present theory and has established the importance of the microscopic nonlocal response theory.

The author is grateful to his coworkers, Prof. H. Ishihara, Prof. Y. Ohfuti, Dr. H. Ajiki, and many graduate students for their collaboration in constructing this framework and examining its potential in various applications. He also acknowledges the long-standing influence of Prof. Y. Toyozawa, which has helped the author in choosing problems and building more consistent theories. Several important pieces of material in this book have been produced since 1998 from the activities of the author's group in the 'center of excellence' project (10CE2004) of the Japanese Ministry of Education, Culture, Sports, Science and Technology. Finally, the author expresses his sincere thanks to his wife Satsuki for her continual support, based on her cheerful and positive attitude to life.

Kobe, Japan *Kikuo Cho*
January 2003

Contents

List of Abbreviations

ABC	additional boundary condition
CM	center of mass
1D	one-dimensional
2D	two-dimensional
3D	three-dimensional
DFWM	degenerate four-wave mixing
ED	electric dipole
e–h	electron–hole
EM	electromagnetic
FWHM	full width at half maximum
FWM	four-wave mixing
L	longitudinal
LP	lower polariton
LWA	long wavelength approximation
MBC	Maxwell boundary condition
MC	micro cavity
MS	mesoscopic
NIDORES	nonlocality induced double resonance in energy and size
NS	nanoscale
PW	plane wave
QD	quantum dot
QED	quantum electrodynamics
Q-factor	quality factor
QW	quantum well
RWA	rotating wave approximation
SHG	second harmonic generation
SNOM	scanning near-field optical microscopy (or microscope)
SR	superradiant
SS	self-sustaining
TE	transverse electric
TM	transverse magnetic
T	transverse
UP	upper polariton
WG	whispering gallery

1. Introduction

1.1 Frameworks for Optical Response Theory

It is well known that the study of radiation–matter interactions has played an important role in the development of modern physics. In particular, quantum mechanics was largely developed from considerations of spectroscopic problems with atoms and blackbodies. Through the spectroscopic study of matter, we have accumulated knowledge about quantum mechanical level structures, transition probabilities, energy transfers, relaxation mechanisms, and so on. Such knowledge has led to the invention of the laser, which has opened a vast new arena of physical research, and the discovery of various kinds of nonlinear optical phenomena, ultra-short optical processes, ultra-high resolution spectroscopy including single-molecule detection, laser cooling, radiation–matter interactions in very high power regimes, etc. On the other hand, synchrotron radiation sources have been developed over a wide frequency range from infrared to gamma rays, with high intensity, well defined pulse profiles, and high brilliance. This allows a wide variety of applications in various fields of science. Today, we can even perform Mössbauer spectroscopy by gamma rays from a synchrotron radiation source and Raman scattering with highly monochromatised syncrotron radiation. The diversity of these characteristic light sources now contributes in essential ways to the high level of optical studies of matter.

The theories supporting the various optical studies may be classified into several groups depending on the energy region, the type of optical process, and the type of matter system to be treated. In each case, we have to determine the motions of charged particles (matter) and the electromagnetic (EM) field. The motion of matter is governed by mechanics (relativistic, or non-relativistic, quantum or classical), whilst the motion of the EM field is determined by Maxwell's equations in the classical case or quantized fields. Various combinations of the two correspond to different theories of the radiation–matter interaction.

Confining ourselves to the non-relativistic cases, we still need to distinguish the quantized and classical EM field. The combination of the classical treatment of the EM field with the quantum mechanical description of matter is said to be semiclassical, and has been widely used with much success in many cases.

However, the semiclassical theory must be further divided into two parts, namely, the macroscopic and the microscopic, according to the nature of the Maxwell equations to be used. Microscopic Maxwell equations describe the microscopic EM field produced by microscopic charge and current densities. To describe the response from a macroscopic body, it is customary to rewrite the microscopic Maxwell equations as macroscopic equations, where the field variables are macroscopically averaged quantities. This averaging process is usually carried out over a 'macroscopically small but microscopically large volume'.

The standard form of the macroscopic semiclassical theory goes as follows. To each different piece of matter in the system one ascribes linear and nonlinear susceptibilities, and in each region of a given susceptibility one solves the macroscopic Maxwell equations. In accordance with the macroscopic nature of the EM field, the susceptibilities are assumed to be local, defining induced polarizations at the same position as the applied field. The solutions of Maxwell's equations in neighboring regions are connected with each other by the (Maxwell) boundary conditions, such as the continuity of tangential or normal components of fields across the interface. Following this procedure for a given incident field, one obtains the response field as a function of position and time.

To describe the EM response from condensed matter, the local macroscopic theory mentioned above has been quite successful, and it is regarded as the 'standard' semiclassical response theory. Historically, the need for a microscopic theory was held in doubt [1], because in the early days, meaningful samples for physical study were all macroscopic.

The appearance of 'mesoscopic' (MS) or 'nanoscale' (NS) systems as reproducible objects for physical study brought awareness of a long-neglected aspect of optical response, i.e., its microscopic and nonlocal character. The size- and shape-dependent response of mesoscopic systems is due to the presence of size-quantized levels which should be described microscopically (quantum mechanically). This leads to a microscopic structure of induced polarization, which is the source of an induced EM field, also to be described microscopically.

Nonlocality appears as a straightforward consequence of a quantum mechanical calculation of susceptibilities [2]. The applied field $\boldsymbol{E}(\boldsymbol{r})$ at a point $\boldsymbol{r}$ induces polarization $\boldsymbol{P}(\boldsymbol{r})$, not only at the same position, but also at other positions within the extent of the relevant wave functions. Therefore, the relationship between $\boldsymbol{E}(\boldsymbol{r})$ and $\boldsymbol{P}(\boldsymbol{r})$ is not local, as for example in a relation of the type $\boldsymbol{P}(\boldsymbol{r}) = \chi(\omega)\boldsymbol{E}(\boldsymbol{r})$. Rather, $\boldsymbol{P}$ is given as a functional of $\boldsymbol{E}(\boldsymbol{r})$.

In MS or NS systems, we often encounter quantized states extended over the whole volume of a sample. The quantization condition is determined by the boundary condition for the wave functions, and the nonlocality of the susceptibilities is also a direct consequence of the coherent spatial structure of the wave functions. Although the size quantization in energy is routinely considered, little attention has been paid to the nonlocality in susceptibilities, in spite of the fact that they have the same origin, i.e., the coherence of wave functions.

It is therefore absolutely necessary in the study of MS or NS systems to use a theoretical approach in which the coherence properties of wave functions are fully retained as required by quantum mechanics. Of course, QED is such a framework, and is in general superior to semiclassical theories. However, its proper use requires the preparation of a complete orthonormal set of states for both matter and field, and rather complicated mathematics including noncommutative field variables. It is also well known that many physical processes can be treated equally well by QED and semiclassical theory, if we do not require the details of the photon ensemble for each mode. Therefore, it is worth trying to construct a new semiclassical theory from the microscopic nonlocal point of view.

The main purpose of this monograph is to describe the framework for such a semiclassical theory of optical response with various applications, and to demonstrate that many optical processes are described from a unified point of view, where the radiative interaction among induced polarizations plays an essential role. (The imaginary part of this interaction is the radiative width of a matter excitation, which has the same analytical form as in QED.)

Nonrelativistic QED and the semiclassical theory of optical response are distinguished by whether they treat the EM field as a quantized or an unquantized quantity, respectively. In the form of a macroscopic local theory, the semiclassical optical response framework moves too far from QED, because all details of the quantum mechanical coherence of matter are smeared out when we take the macroscopic average of field and polarization. Consequently, it is often impossible to make a detailed comparison between them. However, if we adopt the microscopic nonlocal framework as an improved version of the semiclassical theory, it is possible to make a comparison between QED and semiclassical theory, as will be discussed in Sect. 3.7. There, another stage of semiclassical theory is suggested in terms of transition polarizability, introduced to describe Raman and luminescence processes [3].

1.2 Inseparability of Electromagnetism and Mechanics

Whenever we consider the electromagnetic response of matter, we need to determine

- the motion of charged particles (or spins) in the matter induced by the incident EM field,
- the corresponding amplitude of the EM field as a response.

Since they affect one another, they must be determined self-consistently.

The self-consistency of the motions of electrons and the EM field was already noted by Lorentz in his theory of electrons [4]. However, he combined the Maxwell equations with classical Newton mechanics, so that the resulting equations for determining the self-consistent motions constituted a formidable set of nonlinear equations. The correct concept of self-consistency seems therefore to have been blocked by practical difficulties. However, if one uses quantum mechanics instead of Newton mechanics, the prospects for a perturbational treatment look better.

In a semiclassical treatment, the motion of the EM field is determined by Maxwell equations for given charge and current densities, and the motion of charged particles is determined by quantum mechanics in a given EM field (vector and scalar potentials). From the solutions of these equations, we can derive two sets of functional equations relating the vector potential and current density. If we solve them as simultaneous equations, we can determine the self-consistent motions of the EM field and the charged particles. The details will be given in the next chapter.

The inseparability of the motions of the EM field and charged particles is quite obvious in the formulation of QED. From the Hamiltonian for the total system, one derives the Heisenberg equations for the operators representing the EM field and the current density of the matter system. The equations are coupled through the interaction part of the Hamiltonian. It is interesting to compare this set of equations with that of the semiclassical formulation mentioned above. This comparison will be made in Sect. 3.7, which demonstrates the close similarity between the microscopic nonlocal theory and (nonrelativistic) QED.

It should be stressed that the EM field and induced polarization (or current density) to be determined are self-consistent with one another even in the framework of macroscopic local theory, because one solves the coupled equations for these variables. This is self-consistency at the level of the local, macroscopic response. Although it is less sophisticated than at the level of the microscopic, nonlocal response, it is still a type of self-consistency, which can lead, for example, to polaritons (without spatial dispersion), an eigenmode of the coupled radiation–matter system. When the EM field frequency does not coincide with the resonant energies of the matter, self-consistency will not be important, and it will be enough to solve the coupled equations iteratively. In any case, the standard framework of the macroscopic local response gives a self-consistent solution for the EM field and induced polarization on its own level.

1.3 Nonlocality in the Radiation–Matter Interaction

Quite generally, an 'external force' applied at a spacetime point $(\boldsymbol{r}, t)$ will induce a change in the matter system, not only at the same point $(\boldsymbol{r}, t)$, but also at other points $\{(\boldsymbol{r}', t')\}$. The occurrence of the change for $t' > t$ and not for $t' < t$ is called the causality principle. The induced change at $\boldsymbol{r}'$ different from $\boldsymbol{r}$ is called the nonlocal response. If the external force is applied to all the points with varying amplitude and phase, the induced change at $\boldsymbol{r}$ is written as a convolution of the external force and the corresponding 'susceptibility' as an integral kernel. The general occurrence of a nonlocal response may be understood from a simple analogy. A force applied to an elastic string at a point will induce distortion at all the other points. This analogy leads us to the 'extended Lorentz picture of matter' to be discussed in the next section.

The origin of nonlocality lies in the coupling between the dynamical variables associated with different spatial points, which leads to the coherent extension of the eigenmodes or eigenstates of the system. The degree of nonlocality is determined by the spatial extent of the relevant eigenstates, and the contribution of a given eigenstate is determined by the resonance condition for the frequency of the external force with the eigenfrequency of that eigenstate.

As will be shown explicitly in the next chapter, nonlocality follows directly from the quantum mechanical calculation of linear and nonlinear susceptibilities. The fact that the local form of the susceptibility has often been used in traditional optical response theories is closely related to the intention of describing the EM field as macroscopic quantity, i.e., as a quantity averaged over a 'macroscopically small but microscopically large volume'. When the resonant states are well localized, or if the frequency under consideration is well off resonant, the local response can be a good approximation. These are the conditions for the validity of local response theories. If the spatial extent of resonant states is larger than or comparable to the wavelength of the light under consideration, the basis for the local approximation to the susceptibility is no longer valid. A microscopic nonlocal treatment will then be required.

It should be stressed that the difference between the local and nonlocal responses is simply a matter of approximation. In both frameworks, the response field is obtained as a result of self-consistent motions of the EM field and matter (or charged particles). This self-consistency is required to calculate the vector potential and current density as c-numbers in the semiclassical theory. The local approximation is a partial simplification in the self-consistent equations. If the approximation is valid, it should give the same result as the nonlocal treatment.

Historically, the local response framework has been applied to various condensed matter systems and gases with much success, so that most textbooks are written along that line. In that sense, it has become the standard theory of optical response. It appeared quite generally successful, but its major defect was also noticed long ago as the additional boundary condition (ABC) problem for exciton polaritons [5–8] and the EM field near the plasma edge in metal optics [9]. In both cases, standard theory could not derive the condition to fix the ratio of the two (or more) waves in matter for a given frequency of incident light, so that ABC was required in addition to the Maxwell boundary conditions (continuity of field components perpendicular or transverse to the surface, etc.) to provide a unique determination of the field on both sides of the boundary.

As the above example shows, some researchers were aware of the fact that the standard theory contains a fundamental defect. However, the ABC problem was always regarded as a special problem until the emergence of mesoscopic physics in recent times. Once the importance of quantum mechanical coherence, i.e., nonlocality, had been noticed in such mesoscopic systems, it was a straightforward matter to realize its equivalence to the spatial dispersion effect, a central concept in the ABC problem. Since the microscopic approach was known at the time to be the final answer to the ABC problem, it was a natural step to generalize the microscopic approach to arbitrary sizes and shapes of matter. This ends up producing the general microscopic nonlocal framework for optical response. The ABC problem will be discussed in Sect. 3.8 in connection with the development of microscopic nonlocal theory.

1.4 Choice of Matter Hamiltonian and Radiation–Matter Interaction

In this section, we prepare for later discussions by giving some details of the fundamental equations and principal physical quantities in two systems of units. At the same time we describe the two possible ways of defining matter and the EM field, where the longitudinal component of the EM field plays an essential role. In order to make the point clear from first principles, we start the discussion from a fundamental level. As for the units in electromagnetism, we give the expressions in both cgs–Gauss and SI units in this section. Those in curly brackets, viz., $\{\ldots\}_{\mathrm{SI}}$, are in SI units. In the rest of this book, cgs–Gauss units are used, because there are still many papers and books written in this system of units, apart from the classics.

As a general starting point for describing a radiation–matter system, we may take the Lagrangean of the charged particles in given scalar (ϕ) and vector ($\boldsymbol{A}$) potentials and the electromagnetic field in vacuum [10],

$$L = \sum_{\ell} \left[\frac{1}{2} m_\ell v_\ell^2 - e_\ell \phi(\boldsymbol{r}_\ell) + \frac{e_\ell}{c} \boldsymbol{v}_\ell \cdot \boldsymbol{A}(\boldsymbol{r}_\ell) \right] + \int \mathrm{d}\boldsymbol{r}\ \mathcal{L}_{\mathrm{EM}} \tag{1.1}$$

$$\left\{ = \sum_{\ell} \left[\frac{1}{2} m_\ell v_\ell^2 - e_\ell \phi(\boldsymbol{r}_\ell) + e_\ell \boldsymbol{v}_\ell \cdot \boldsymbol{A}(\boldsymbol{r}_\ell) \right] + \int \mathrm{d}\boldsymbol{r} \mathcal{L}_{\mathrm{EM}} \right\}_{\mathrm{SI}} , \tag{1.2}$$

$$\mathcal{L}_{\mathrm{EM}} = \frac{1}{8\pi} \left[\left(\frac{1}{c} \frac{\partial \boldsymbol{A}}{\partial t} + \nabla \phi \right)^2 - (\nabla \times \boldsymbol{A})^2 \right] = \frac{1}{8\pi} (\boldsymbol{E}^2 - \boldsymbol{B}^2) \tag{1.3}$$

$$\left\{ = \frac{\epsilon_0}{2} \left[\left(\frac{\partial \boldsymbol{A}}{\partial t} + \nabla \phi \right)^2 - c^2 (\nabla \times \boldsymbol{A})^2 \right] = \frac{\epsilon_0}{2} (\boldsymbol{E}^2 - c^2 \boldsymbol{B}^2) \right\}_{\mathrm{SI}} . \tag{1.4}$$

Generalized coordinates in this Lagrangean are $\{\boldsymbol{r}_\ell, \phi(\boldsymbol{r}), \boldsymbol{A}(\boldsymbol{r})\}$, and the time derivative of ϕ does not appear. Let us take the Coulomb gauge $\nabla\cdot\boldsymbol{A} = 0$. Then the longitudinal (L) and transverse (T) parts of the electric field are $-\nabla\phi$ and $(-1/c)\partial\boldsymbol{A}/\partial t$, respectively. The principle of least action leads to the three types of Lagrange equations:
(A) the equation of motion for each charged particle under the Lorentz force,

$$m\frac{\mathrm{d}\boldsymbol{v}_\ell}{\mathrm{d}t} = e_\ell\left[-\nabla\phi(\boldsymbol{r}_\ell) - \frac{1}{c}\frac{\partial\boldsymbol{A}(\boldsymbol{r}_\ell)}{\partial t} + \frac{\boldsymbol{v}_\ell}{c}\times[\nabla\times\boldsymbol{A}(\boldsymbol{r}_\ell)]\right] \tag{1.5}$$

$$= e_\ell\left(\boldsymbol{E} + \frac{\boldsymbol{v}_\ell}{c}\times\boldsymbol{B}\right) \tag{1.6}$$

$$\left\{= e_\ell\left[-\nabla\phi(\boldsymbol{r}_\ell) - \frac{\partial\boldsymbol{A}(\boldsymbol{r}_\ell)}{\partial t} + \boldsymbol{v}_\ell\times\nabla\times\boldsymbol{A}(\boldsymbol{r}_\ell)\right]\right\}_{\mathrm{SI}} \tag{1.7}$$

$$\left\{= e_\ell(\boldsymbol{E} + \boldsymbol{v}_\ell\times\boldsymbol{B})\right\}_{\mathrm{SI}}, \tag{1.8}$$

(B) the Poisson equation for the scalar potential,

$$\nabla^2\phi = -4\pi\rho \quad \left\{= -\frac{1}{\epsilon_0}\rho\right\}_{\mathrm{SI}}, \tag{1.9}$$

and (C) the equation of motion for the vector potential,

$$\frac{1}{c^2}\frac{\partial^2\boldsymbol{A}}{\partial t^2} - \nabla^2\boldsymbol{A} = \frac{4\pi}{c}\boldsymbol{J}(\boldsymbol{r}) \quad \left\{= \frac{1}{\epsilon_0 c^2}\boldsymbol{J}(\boldsymbol{r})\right\}_{\mathrm{SI}}, \tag{1.10}$$

where the charge density $\rho(\boldsymbol{r})$ and current density $\boldsymbol{J}(\boldsymbol{r})$ are defined as

$$\rho(\boldsymbol{r}) = \sum_\ell e_\ell\,\delta(\boldsymbol{r} - \boldsymbol{r}_\ell), \tag{1.11}$$

$$\boldsymbol{J}(\boldsymbol{r}) = \sum_\ell e_\ell\,\boldsymbol{v}_\ell\,\delta(\boldsymbol{r} - \boldsymbol{r}_\ell). \tag{1.12}$$

In terms of ρ and $\boldsymbol{J}$, the interaction of the charged particles and the EM field can be rewritten in integral form as

$$\sum_\ell\left[-e_\ell\phi(\boldsymbol{r}_\ell) + \frac{e_\ell}{c}\boldsymbol{v}_\ell\cdot\boldsymbol{A}(\boldsymbol{r}_\ell)\right] = \int\mathrm{d}\boldsymbol{r}\left[-\rho(\boldsymbol{r})\phi(\boldsymbol{r}) + \frac{1}{c}\boldsymbol{J}(\boldsymbol{r})\cdot\boldsymbol{A}(\boldsymbol{r})\right], \tag{1.13}$$

$$\left\{\sum_\ell[-e_\ell\phi(\boldsymbol{r}_\ell) + e_\ell\boldsymbol{v}_\ell\cdot\boldsymbol{A}(\boldsymbol{r}_\ell)] = \int\mathrm{d}\boldsymbol{r}\,[-\rho(\boldsymbol{r})\phi(\boldsymbol{r}) + \boldsymbol{J}(\boldsymbol{r})\cdot\boldsymbol{A}(\boldsymbol{r})]\right\}_{\mathrm{SI}}. \tag{1.14}$$

In terms of the solution to the Poisson equation

$$\phi(\boldsymbol{r}) = \sum_\ell\frac{e_\ell}{|\boldsymbol{r} - \boldsymbol{r}_\ell|} \quad \left\{= \sum_\ell\frac{e_\ell}{4\pi\epsilon_0|\boldsymbol{r} - \boldsymbol{r}_\ell|}\right\}_{\mathrm{SI}}, \tag{1.15}$$

one can show that the sum U_{s} of the L field energy and the interaction between the charged particles and the L field turns out to be the Coulomb interaction among the charged particles, viz.,

$$U_{\mathrm{s}} = \int \mathrm{d}\boldsymbol{r} \left[\frac{1}{8\pi} (\nabla\phi)^2 - \rho(\boldsymbol{r})\phi(\boldsymbol{r}) \right] = -\frac{1}{2} \sum_{\ell \neq \ell'} \frac{e_\ell e_{\ell'}}{|\boldsymbol{r}_\ell - \boldsymbol{r}_{\ell'}|} \tag{1.16}$$

$$\left\{ = \int \mathrm{d}\boldsymbol{r} \left[\frac{\epsilon_0}{2} (\nabla\phi)^2 - \rho(\boldsymbol{r})\phi(\boldsymbol{r}) \right] = -\frac{1}{2} \sum_{\ell \neq \ell'} \frac{e_\ell e_{\ell'}}{4\pi\epsilon_0 |\boldsymbol{r}_\ell - \boldsymbol{r}_{\ell'}|} \right\}_{\mathrm{SI}} , \tag{1.17}$$

where the self-energy of each charged particle in the potential produced by itself ($\ell = \ell'$ terms) is neglected. The Hamiltonian of the charged particles is

$$H_{\mathrm{M}} = \sum_\ell \frac{1}{2m_\ell} \left[\boldsymbol{p}_\ell - \frac{e_\ell}{c} \boldsymbol{A}(\boldsymbol{r}_\ell) \right]^2 + \frac{1}{2} \sum_{\ell \neq \ell'} \frac{e_\ell e_{\ell'}}{|\boldsymbol{r}_\ell - \boldsymbol{r}_{\ell'}|} \tag{1.18}$$

$$\left\{ = \sum_\ell \frac{1}{2m_\ell} \left[\boldsymbol{p}_\ell - e_\ell \boldsymbol{A}(\boldsymbol{r}_\ell) \right]^2 + \frac{1}{2} \sum_{\ell \neq \ell'} \frac{e_\ell e_{\ell'}}{4\pi\epsilon_0 |\boldsymbol{r}_\ell - \boldsymbol{r}_{\ell'}|} \right\}_{\mathrm{SI}} . \tag{1.19}$$

Adding the Hamiltonian of the transverse EM field to the above, we obtain the full, general Hamiltonian for the interacting radiation–matter system.

Since we want to construct a first-principles theory, it is important to define the matter Hamiltonian and EM field precisely. There are essentially two choices. The first is to adopt the above-mentioned scheme, where the EM field is just the pure T component and the full Coulomb interaction among charged particles is included in the matter Hamiltonian. In this case, the susceptibility of induced polarization is defined with respect to the T components of the Maxwell field. The induced polarization can have L components as well as T components, and this is the source of an L field

$$\boldsymbol{E}_{\mathrm{L}}(\boldsymbol{r}, \omega) = \nabla \int \mathrm{d}\boldsymbol{r}' \, \frac{\nabla \cdot \boldsymbol{P}(\boldsymbol{r}', \omega)}{|\boldsymbol{r} - \boldsymbol{r}'|} \quad \left\{ = \nabla \int \mathrm{d}\boldsymbol{r}' \, \frac{\nabla \cdot \boldsymbol{P}(\boldsymbol{r}', \omega)}{4\pi\epsilon_0 |\boldsymbol{r} - \boldsymbol{r}'|} \right\}_{\mathrm{SI}} . \tag{1.20}$$

This L field is the instantaneous Coulomb field due to the induced charge density $-\nabla \cdot \boldsymbol{P}$. Thus, the interaction of this field with charge density in the matter is simply a part of the Coulomb interaction among the charges, treated as part of the internal energy of the matter system. In particular, the interaction with the induced polarization itself produces an additional energy term reflecting the L, T, or LT-mixed character of the induced polarization. In this scheme, the excitation energies of the matter, whatever T, L, or LT-mixed character they may have, always appear as poles of susceptibilities. This should be compared with the second choice to be mentioned below, where the T-mode excitation appears at the pole and the L-mode at the zero of the dielectric function.

The second choice of matter Hamiltonian and EM field corresponds to another, more popular approach to the description of the EM response of matter, where the L component of the induced electric field, $\boldsymbol{E}_{\mathrm{L}}$ in (1.21), is regarded as part of the 'external' field acting on the 'matter'. This choice affects the form of the matter Hamiltonian. Obviously we cannot maintain the full Coulomb interaction in the matter Hamiltonian in this scheme, because otherwise we would double count a certain part of the Coulomb interaction. It has been pointed out that a certain part of the Coulomb interaction should be omitted from the matter Hamiltonian in order to be consistent [11]. The part to be omitted is the Coulomb interaction between polarization charges, or equivalently, the interaction energy between the L field and the polarization:

$$H_{\mathrm{C}} = -\int \mathrm{d}\boldsymbol{r}\, \boldsymbol{E}_{\mathrm{L}}(\boldsymbol{r},\omega)\cdot\boldsymbol{P}(\boldsymbol{r},\omega) \tag{1.21}$$

$$= \int \mathrm{d}\boldsymbol{r} \int \mathrm{d}\boldsymbol{r}' \frac{\nabla\cdot\boldsymbol{P}(\boldsymbol{r},\omega)\nabla'\cdot\boldsymbol{P}(\boldsymbol{r}',\omega)}{|\boldsymbol{r}-\boldsymbol{r}'|} \tag{1.22}$$

$$\left\{ = \int \mathrm{d}\boldsymbol{r} \int \mathrm{d}\boldsymbol{r}' \frac{\nabla\cdot\boldsymbol{P}(\boldsymbol{r},\omega)\nabla'\cdot\boldsymbol{P}(\boldsymbol{r}',\omega)}{4\pi\epsilon_0|\boldsymbol{r}-\boldsymbol{r}'|} \right\}_{\mathrm{SI}} . \tag{1.23}$$

The problem of choosing the external field and corresponding matter Hamiltonian has been studied in the case of excitons in 3D crystals [12], and it was shown that the relevant part of the matter Hamiltonian to be subtracted is the electron–hole (e–h) exchange interaction H_{exch}. It is interesting to note that the e–h exchange interaction in a system of bulk excitons is a special case of H_{C} [13]. The expression for H_{C} applies to more general situations, where the origin of polarization may be either electronic or lattice vibrational, and the system may or may not be confined. In the case of a bulk material, this energy causes the energy splitting between the T and L polarization waves of excitons or optical phonons. This splitting may be described as the energy difference between matter eigenstates, or as being due to the difference in the coupling with the depolarization field.

The two choices of matter Hamiltonian and external EM field require the corresponding susceptibilities, one with respect to the full Maxwell field and the other with respect to its T component alone. The difference between the two Hamiltonians H_{C} is chosen so that the physical quantities obtained from these two susceptibilities are equivalent [12, 14].

Throughout most of this book, we make the first choice, in which the matter Hamiltonian contains the full Coulomb interaction and the EM field is purely transverse. This is because the equivalence of the two choices has been shown only in the linear response case, and it is safer to stick with the correct scheme, i.e., the first choice. In some linear response problems, it is easier to work with the second choice, and these are mentioned in Sects. 3.3 and 4.1.3.

1.5 Extended Lorentz Picture for Interacting Radiation–Matter Systems

Once we have chosen the matter Hamiltonian containing the full Coulomb interaction among charged particles interacting with the transverse EM field, we are ready to start the self-consistent determination of the microscopic motions of the matter and EM field. As mentioned before, the macroscopic response theory needs to be modified to include nonlocal features arising from the microscopic quantum mechanical calculation. The details of such a framework will be given in the next chapter, but here we will describe the physical picture underlying this approach.

The matter system is represented by the induced current and charge densities in the Maxwell equations. The charge density is related to the longitudinal part of the current density through the continuity equation. It is therefore enough to consider the current density with both longitudinal and transverse parts. The current density consists of various components with characteristic frequencies corresponding to the transition energies of the matter system. In the absence of the EM field, they are just independent oscillators, with possible damping effects.

In the presence of an EM field, there are two effects on the current density. Firstly, oscillations are induced in each of its components, with an amplitude depending on the frequency detuning, the coupling strength and the damping. The induced oscillation of the current density emits an EM field, which further interacts with various components of the current density. Thus the second effect is the exchange of (transverse) EM field among the components of the current density.

According to the classical Lorentz model, a matter system behaves like an assembly of electric oscillators. This model provides a good description for the frequency dependence of the optical response of the matter near a resonance. It has the same functional form as the macroscopic susceptibility obtained from a (macroscopically averaged) quantum mechanical calculation.

It is interesting that the physical picture of an interacting radiation–matter system described by the microscopic nonlocal theory may be viewed as an 'extended Lorentz picture'. In particular, a matter system is an assembly of oscillating current densities with characteristic eigenfrequencies. Although such a description is a 'fact' derived from a first-principles theory rather than a 'model', it represents the same physical picture as the original Lorentz model.

If we compare the extended Lorentz picture with the original Lorentz model, there are two generalized aspects:

- Each component of the current density has a coherent spatial extension which could be larger than the wavelength of the interacting EM field. This requires a nonlocal description of the optical response, which leads

to a self-consistent determination of the spatial variations of the current density and EM field.
- Various current densities interact with one another via the exchange of transverse EM field.

The second aspect is an essential ingredient of this formalism. Each component of the current density is oscillating in time, and it emits a transverse EM field according to the Maxwell equations. The emitted EM field propagates with the velocity of light and interacts with other components of the current density. Such a process may be called the retarded interaction between current densities. The interaction energy due to this process is written as the interaction of one component of current density with the transverse field produced by the other. The details will be given in Chap. 2. Including this effect leads to a correct description of the size, shape, and internal structure dependence of the resonant optical responses of MS or NS systems. The eigenmode of the coupled radiation–matter system giving the resonant structure of the response spectrum is obtained by diagonalizing the matter excitation energy and retarded interaction. Examples of such an eigenmode can be found in any coupled system, such as polaritons, dynamically scattered X-rays, surface EM modes, whispering gallery modes in a dielectric sphere, etc., with or without radiative decay.

1.6 Relation with Other Frameworks

Most of the contents of this book have been worked out from the activities of the author and his group, independently of other groups. The purpose of the book is not to write a review of related works, but to shed more light on the important but less popular theoretical framework of microscopic nonlocal response. Its importance is gradually becoming better known, especially in studies of nanostructures, but is still far from sufficient in comparison with the traditional macroscopic (local) response theory.

To the author's knowledge, there are not many groups working in this direction to produce new types of approach. We will briefly mention two types of approach by the groups of Keller [15] and Stahl [16]. They have the same intention as ourselves, namely, to take the coherence of matter-excited states into account in calculating the linear and nonlinear optical responses of various forms of matter within the semiclassical scheme. Keller's approach is rather close to ours in setting up two integral equations for the EM field and transition current density, while Stahl's coherent wave approach uses 'interband transition amplitudes' to set up the constitutive equation between the interband current density and electric field. Due to stronger interests in metals, Keller seems to prefer the EM field with the L component, i.e., the Maxwell field, as the matter-polarizing 'external field', and the matter Hamiltonian that does not contain the full Coulomb interaction. Stahl's method has been used extensively by Bassani's group for various problems related to excitons [17].

Due to the essential similarity between these frameworks and our own, the merits and demerits should be sought in the details of their application, the validity limit of the approximations involved, and the generality of their formulation. It is not yet appropriate to make such a comparison, but it is hoped that someone will do so in the near future.

2. Formulation of Nonlocal Response Theory

2.1 Microscopic Maxwell Equations

Since we intend to describe mesoscopic systems as in the atomic case, maintaining full quantum mechanical coherence, we must carry out a microscopic treatment of both the matter and the EM field. In contrast to the prevailing optical response theory of condensed matter, where the details of the quantum mechanical motion of matter and the microscopic variation of the EM field are smeared out, we keep those details required from first principles.

In the microscopic Maxwell equations there are not four field variables $\boldsymbol{E}, \boldsymbol{D}, \boldsymbol{H}, \boldsymbol{B}$, but only two, and the charge ρ and current density $\boldsymbol{j}$ are microscopic quantities. Their explicit forms in terms of vector and scalar potentials, $\boldsymbol{A}$ and ϕ, respectively, are given by [18]

$$\frac{1}{c^2}\frac{\partial^2 \boldsymbol{A}}{\partial t^2} - \nabla^2 \boldsymbol{A} + \nabla\left(\nabla\cdot\boldsymbol{A} + \frac{1}{c}\frac{\partial \phi}{\partial t}\right) = \frac{4\pi}{c}\,\boldsymbol{j}\,, \tag{2.1}$$

$$-\nabla^2\phi - \frac{1}{c}\nabla\cdot\frac{\partial \boldsymbol{A}}{\partial t} = 4\pi\rho\,, \tag{2.2}$$

where c is the velocity of light in vacuum. The electric and magnetic fields, $\boldsymbol{E}$ and $\boldsymbol{B}$, respectively, are given by

$$\boldsymbol{E} = -\nabla\phi - \frac{1}{c}\frac{\partial \boldsymbol{A}}{\partial t}\,, \tag{2.3}$$

$$\boldsymbol{B} = \nabla\times\boldsymbol{A}\,. \tag{2.4}$$

The charge and current densities are related via the continuity equation

$$\nabla\cdot\boldsymbol{j} + \frac{\partial \rho}{\partial t} = 0\,, \tag{2.5}$$

as seen by direct differentiation of (2.1) and (2.2). Choosing the Coulomb gauge,

$$\nabla\cdot\boldsymbol{A} = 0\,, \tag{2.6}$$

we can immediately solve (2.2) to obtain

$$\phi(\boldsymbol{r},t) = \int \mathrm{d}\boldsymbol{r}'\,\frac{1}{|\boldsymbol{r}-\boldsymbol{r}'|}\,\rho(\boldsymbol{r}',t)\,. \tag{2.7}$$

In terms of this result and the continuity equation (2.5), the ωth Fourier component of (2.1) can be rewritten as

$$-(\nabla^2 + q^2)\tilde{\boldsymbol{A}}(\boldsymbol{r},\omega) = \frac{4\pi}{c}\tilde{\boldsymbol{j}}(\boldsymbol{r},\omega) + \frac{1}{c}\int d\boldsymbol{r}'\,\frac{\nabla'\nabla'\cdot\tilde{\boldsymbol{j}}(\boldsymbol{r}',\omega)}{|\boldsymbol{r}-\boldsymbol{r}'|}\,, \tag{2.8}$$

where a partial integration has been carried out in the integral and q is the wave number in vacuum, viz.,

$$q = \frac{\omega}{c}\,. \tag{2.9}$$

Defining the Green function G_q satisfying the equation

$$\left(\nabla^2 + q^2\right) G_q(\boldsymbol{r},\omega) = -4\pi\delta(\boldsymbol{r})\,, \tag{2.10}$$

we can easily write down the solution of (2.8) as

$$\begin{aligned}\tilde{\boldsymbol{A}}(\boldsymbol{r}) = \tilde{\boldsymbol{A}}_0(\boldsymbol{r}) &+ \frac{1}{c}\int d\boldsymbol{r}'\, G_q(\boldsymbol{r}-\boldsymbol{r}')\,\tilde{\boldsymbol{j}}(\boldsymbol{r}') \\ &+ \frac{1}{4\pi c}\int d\boldsymbol{r}'\int d\boldsymbol{r}''\, G_q(\boldsymbol{r}-\boldsymbol{r}')\frac{\nabla''\nabla''\cdot\tilde{\boldsymbol{j}}(\boldsymbol{r}'')}{|\boldsymbol{r}'-\boldsymbol{r}''|}\,,\end{aligned} \tag{2.11}$$

where the argument ω is omitted for simplicity and $\boldsymbol{A}_0$ is the free field, usually corresponding to an incident field. Rewriting the Green function in the double integral in terms of (2.10) and performing partial integration twice, we get the final expression for the field as

$$\tilde{\boldsymbol{A}}(\boldsymbol{r}) = \tilde{\boldsymbol{A}}_0 + \mathcal{G}[\tilde{\boldsymbol{j}}]\,, \tag{2.12}$$

$$\mathcal{G}[\tilde{\boldsymbol{j}}] = \frac{1}{c}\int d\boldsymbol{r}'\,\mathbf{G}^{(\mathrm{T})}(\boldsymbol{r}-\boldsymbol{r}')\cdot\tilde{\boldsymbol{j}}(\boldsymbol{r}')\,, \tag{2.13}$$

$$\mathbf{G}^{(\mathrm{T})}(\boldsymbol{r}-\boldsymbol{r}') = G_q(\boldsymbol{r}-\boldsymbol{r}')\mathbf{1} + \frac{1}{q^2}\left[G_q(\boldsymbol{r}-\boldsymbol{r}') - G_0(\boldsymbol{r}-\boldsymbol{r}')\right]\nabla'\nabla'\,. \tag{2.14}$$

Equation (2.13) defines the form of the functional $\mathcal{G}$. This result shows that the response field is the sum of the incident field and the induced field caused by the current density. It should be noted that the tensorial Green function $\mathbf{G}^{(\mathrm{T})}$ produces a transverse field at $\boldsymbol{r}$ from the current density at $\boldsymbol{r}'$. Another expression for the same field is

$$\mathcal{G}[\tilde{\boldsymbol{j}}] = \frac{1}{c}\int d\boldsymbol{r}'\, G_q(\boldsymbol{r}-\boldsymbol{r}')\tilde{\boldsymbol{j}}^{(\mathrm{T})}(\boldsymbol{r}')\,, \tag{2.15}$$

where $\tilde{\boldsymbol{j}}^{(\mathrm{T})}(\boldsymbol{r})$ is the transverse component of the current density defined in terms of the Fourier transform $\tilde{\boldsymbol{j}}(\boldsymbol{k},\omega)$ of $\tilde{\boldsymbol{j}}(\boldsymbol{r},\omega)$ as

$$\tilde{\boldsymbol{j}}^{(\mathrm{T})}(\boldsymbol{r}) = \frac{V}{8\pi^3}\int d\boldsymbol{k}\, \mathrm{e}^{\mathrm{i}\boldsymbol{k}\cdot\boldsymbol{r}}\left[\mathbf{1} - \hat{\boldsymbol{e}}_3(\boldsymbol{k})\hat{\boldsymbol{e}}_3(\boldsymbol{k})\right]\cdot\tilde{\boldsymbol{j}}(\boldsymbol{k})\,, \tag{2.16}$$

where $\hat{\boldsymbol{e}}_3(\boldsymbol{k})$ is the unit vector in the direction of $\boldsymbol{k}$.

The boundary condition for the Green function G_q is chosen in such a way that the response propagates with time from the matter to infinity, i.e.,

$$G_q(\boldsymbol{r}) = \frac{\mathrm{e}^{\mathrm{i}q|\boldsymbol{r}|}}{|\boldsymbol{r}|} \ . \tag{2.17}$$

The outgoing character of this solution is obvious, if we note that the ω th Fourier component is associated with the phase factor $\exp(-\mathrm{i}\omega t)$.

It is sometimes required to write (2.12) as a relation between the electric field $\tilde{\boldsymbol{E}}$ and the induced polarization $\tilde{\boldsymbol{P}}$. For this purpose, we define $\tilde{\boldsymbol{P}}$ and the transverse source field $\tilde{\boldsymbol{E}}_\mathrm{s}$ as

$$\tilde{\boldsymbol{j}}(\boldsymbol{r},\omega) = -\mathrm{i}\omega\tilde{\boldsymbol{P}}(\boldsymbol{r},\omega) \ , \tag{2.18}$$

$$\tilde{\boldsymbol{E}}_\mathrm{s}(\boldsymbol{r},\omega) = \mathrm{i}q\tilde{\boldsymbol{A}}(\boldsymbol{r},\omega) \ , \tag{2.19}$$

$$\tilde{\boldsymbol{E}}_0(\boldsymbol{r},\omega) = \mathrm{i}q\tilde{\boldsymbol{A}}_0(\boldsymbol{r},\omega) \ . \tag{2.20}$$

Then (2.12) can be rewritten as

$$\begin{aligned}\tilde{\boldsymbol{E}}_\mathrm{s}(\boldsymbol{r}) &= \tilde{\boldsymbol{E}}_0(\boldsymbol{r}) + \tilde{\boldsymbol{E}}_\mathrm{r}(\boldsymbol{r}) + \frac{1}{q^2}\nabla\nabla\cdot\tilde{\boldsymbol{E}}_\mathrm{r}(\boldsymbol{r}) - \tilde{\boldsymbol{E}}_\mathrm{dep}(\boldsymbol{r}) \ , \\ &= \tilde{\boldsymbol{E}} - \tilde{\boldsymbol{E}}_\mathrm{dep} \ ,\end{aligned} \tag{2.21}$$

where

$$\tilde{\boldsymbol{E}} = \tilde{\boldsymbol{E}}_0 + \tilde{\boldsymbol{E}}_\mathrm{r} + \frac{1}{q^2}\nabla\nabla\cdot\tilde{\boldsymbol{E}}_\mathrm{r} \ , \tag{2.22}$$

$$\tilde{\boldsymbol{E}}_\mathrm{r}(\boldsymbol{r},\omega) = q^2\int \mathrm{d}\boldsymbol{r}'\, G_q(\boldsymbol{r}-\boldsymbol{r}')\tilde{\boldsymbol{P}}(\boldsymbol{r}',\omega) \ , \tag{2.23}$$

$$\tilde{\boldsymbol{E}}_\mathrm{dep}(\boldsymbol{r},\omega) = \int \mathrm{d}\boldsymbol{r}'\, \frac{\nabla'\nabla'\cdot\tilde{\boldsymbol{P}}(\boldsymbol{r},\omega)}{|\boldsymbol{r}-\boldsymbol{r}'|} \ . \tag{2.24}$$

It should be noted that the field $\tilde{\boldsymbol{E}}(\boldsymbol{r},\omega)$, defined by (2.22) in terms of $\tilde{\boldsymbol{E}}_0$ and $\tilde{\boldsymbol{E}}_\mathrm{r}$, satisfies the Maxwell equation

$$\nabla\times(\nabla\times\tilde{\boldsymbol{E}}) - q^2\tilde{\boldsymbol{E}} = 4\pi q^2\tilde{\boldsymbol{P}} \ , \tag{2.25}$$

which turns out to be the following equation for $\tilde{\boldsymbol{E}}_\mathrm{r}$:

$$-\left(\nabla^2 + q^2\right)\tilde{\boldsymbol{E}}_\mathrm{r} = 4\pi q^2\tilde{\boldsymbol{P}} \ . \tag{2.26}$$

Its solution is (2.23). The Maxwell field $\tilde{\boldsymbol{E}}$ is thus divided into the transverse part $\tilde{\boldsymbol{E}}_\mathrm{s}$ and the longitudinal part $\tilde{\boldsymbol{E}}_\mathrm{dep}$. The suffix on the latter stands for 'depolarization'. Rewriting (2.24) as

$$\tilde{\boldsymbol{E}}_\mathrm{dep}(\boldsymbol{r}) = -\nabla\int \mathrm{d}\boldsymbol{r}'\, \frac{-\nabla'\cdot\tilde{\boldsymbol{P}}(\boldsymbol{r}')}{|\boldsymbol{r}-\boldsymbol{r}'|} \ , \tag{2.27}$$

and noting that $-\nabla\cdot\tilde{\boldsymbol{P}}$ represents the charge density induced by polarization, we can see directly that the integral represents the potential due to the induced charge density, and that $\tilde{\boldsymbol{E}}_\mathrm{dep}$ represents the (longitudinal) field induced by polarization. Although the transverse character of $\tilde{\boldsymbol{E}}_{\mathrm{s}}$ is obvious from (2.19) together with (2.6), it can also be derived directly from the definition (2.21):

$$\begin{aligned}
\nabla\cdot\tilde{\boldsymbol{E}}_{\mathrm{s}} &= \nabla\cdot\left(\tilde{\boldsymbol{E}}_0 + \tilde{\boldsymbol{E}}_{\mathrm{r}} + \frac{1}{q^2}\nabla\nabla\cdot\tilde{\boldsymbol{E}}_{\mathrm{r}}\right) - \nabla\cdot\tilde{\boldsymbol{E}}_{\mathrm{dep}} \\
&= \nabla\cdot\tilde{\boldsymbol{E}}_{\mathrm{r}} + \frac{1}{q^2}\nabla\cdot\nabla^2\tilde{\boldsymbol{E}}_{\mathrm{r}} - \nabla^2\int \mathrm{d}\boldsymbol{r}'\,\frac{\nabla'\cdot\tilde{\boldsymbol{P}}(\boldsymbol{r}')}{|\boldsymbol{r}-\boldsymbol{r}'|} \\
&= -4\pi\nabla\cdot\tilde{\boldsymbol{P}} + 4\pi\int \mathrm{d}\boldsymbol{r}'\,\delta(\boldsymbol{r}-\boldsymbol{r}')\nabla'\cdot\tilde{\boldsymbol{P}}(\boldsymbol{r}')\ , \\
&= 0\ .
\end{aligned}$$

In order to determine the current density or induced polarization caused by the incident field, we need to consider the motion of the matter system, using the quantum mechanics of charged particles in a given EM field. This is done in the following section.

2.2 Motion of the Matter System

The general (nonrelativistic) Hamiltonian for an assembly of charged particles in a given transverse EM field was derived in (1.18) as

$$H_{\mathrm{M}} = \sum_{\ell}\frac{1}{2m_\ell}\left[\boldsymbol{p}_\ell - \frac{e_\ell}{c}\boldsymbol{A}(\boldsymbol{r}_\ell)\right]^2 + \frac{1}{2}\sum_{\ell\neq\ell'}\frac{e_\ell e_{\ell'}}{|\boldsymbol{r}_\ell-\boldsymbol{r}_{\ell'}|}\ , \tag{2.28}$$

where $e_\ell, m_\ell, \boldsymbol{r}_\ell$, and $\boldsymbol{p}_\ell$ are the charge, mass, coordinate, and conjugate momentum of the coordinate, respectively, for the ℓth particle. The transverse field is represented by the vector potential $\boldsymbol{A}$ in the Coulomb gauge. In Sect. 1.4, this Hamiltonian is shown to be consistent with the equations of motion of the charged particles under the Lorentz force. Let us divide up the Hamiltonian as

$$H_{\mathrm{M}} = H_0 + H_{\mathrm{int}}\ , \tag{2.29}$$

$$H_0 = \sum_{\ell}\frac{1}{2m_\ell}\boldsymbol{p}_\ell^2 + \frac{1}{2}\sum_{\ell\neq\ell'}\frac{e_\ell e_{\ell'}}{|\boldsymbol{r}_\ell-\boldsymbol{r}_{\ell'}|}\ , \tag{2.30}$$

$$H_{\mathrm{int}} = \sum_{\ell}\left(-\frac{e_\ell}{m_\ell c}\boldsymbol{A}_\ell\cdot\boldsymbol{p}_\ell + \frac{e_\ell^2}{2m_\ell c^2}\boldsymbol{A}_\ell^2\right)\ . \tag{2.31}$$

We regard H_0 as the matter Hamiltonian and H_{int} as the radiation–matter interaction.

Since we do not consider any external charge source, the scalar potential ϕ is produced only by the charged particles in the system itself, i.e.,

$$\phi(\boldsymbol{r},t) = \int \mathrm{d}\boldsymbol{r}'\,\frac{\rho(\boldsymbol{r}',t)}{|\boldsymbol{r}-\boldsymbol{r}'|}\ , \tag{2.32}$$

$$\rho(\boldsymbol{r},t) = \sum_{\ell} e_\ell \delta(\boldsymbol{r} - \boldsymbol{r}_\ell) , \tag{2.33}$$

where ρ is the charge density of the matter. The terms related by the scalar potential in H_0 are rewritten into the Coulomb interaction between the charged particles, as mentioned in Sect. 1.4. Each of the self-interaction terms ($\ell = \ell'$ terms) omitted in the Coulomb interaction is a large constant related to the electron radius [10].

The above-mentioned choice of matter Hamiltonian agrees with the usual picture of matter in the nonrelativistic regime. For example, a hydrogen atom consists of an electron and a proton interacting via the instantaneous Coulomb interaction.

The radiation field in the Coulomb gauge is purely transverse as in (2.6), and the radiation–matter interaction (2.31) contains only the vector potential. Due to the transverse nature of $\boldsymbol{A}$, the order of $\boldsymbol{A}$ and $\boldsymbol{p}$ does not make any difference in (2.31).

The matter state prepared at time $t = t_0$ as an eigenstate of H_0 will experience time evolution after switching on the radiation–matter interaction. This time evolution is described by the Schrödinger equation

$$\mathrm{i}\hbar \frac{\partial \psi}{\partial t} = (H_0 + H_{\mathrm{int}})\psi . \tag{2.34}$$

This equation can be solved perturbatively with respect to H_{int}. We first rewrite the wave function as

$$\psi = \exp(-\mathrm{i}H_0 t/\hbar)\tilde{\psi} \tag{2.35}$$

and then solve the equation for $\tilde{\psi}$

$$\mathrm{i}\hbar \frac{\partial \tilde{\psi}}{\partial t} = H'(t)\tilde{\psi} , \tag{2.36}$$

where

$$H'(t) = \exp(\mathrm{i}H_0 t/\hbar) H_{\mathrm{int}} \exp(-\mathrm{i}H_0 t/\hbar) . \tag{2.37}$$

The result is

$$\tilde{\psi}(t) = \hat{T} \exp\left[-\frac{\mathrm{i}}{\hbar}\int_{t_0}^{t} \mathrm{d}\tau\, H'(\tau)\right] \tilde{\psi}(t_0) \tag{2.38}$$

$$= \tilde{\psi}(t_0) + \left(\frac{-\mathrm{i}}{\hbar}\right)\int_{t_0}^{t} \mathrm{d}\tau_1\, H'(\tau_1)\tilde{\psi}(t_0) + \left(\frac{-\mathrm{i}}{\hbar}\right)^2 \int_{t_0}^{t} \mathrm{d}\tau_1 \int_{t_0}^{\tau_1} \mathrm{d}\tau_2\, H'(\tau_1)H'(\tau_2)\tilde{\psi}(t_0) + \cdots . \tag{2.39}$$

The time-ordering operator $\hat{T}$ may be introduced in (2.38) to give a compact expression for the result.

The expectation value at time t of an operator $\hat{b}$ is given by

$$\langle b \rangle_t = \langle \psi(t)|\hat{b}|\psi(t)\rangle . \tag{2.40}$$

When the initial state of the matter is prepared more generally as an ensemble, it is better to follow the time evolution of the density matrix $\bar{\rho}$ of the matter:

$$i\hbar \frac{d\bar{\rho}}{dt} = [H_0 + H_{\rm int}\,, \bar{\rho}]\ . \tag{2.41}$$

This can be solved as a perturbation expansion with respect to $H_{\rm int}$. The expectation value of the operator $\hat{b}$ is calculated as

$$\langle b \rangle_t = {\rm Tr}\left\{\hat{b}\ \bar{\rho}(t)\right\}\ . \tag{2.42}$$

This result is a generalization of (2.40), in the sense that the initial state of the system is prepared, not as an eigenstate of H_0, but as an ensemble over various states.

The quantity of interest here is the current density. A general way to define its operator form is to start from the charge density operator

$$\hat{n}(\boldsymbol{r}) = \sum_\ell e_\ell \delta(\boldsymbol{r} - \boldsymbol{r}_\ell) \tag{2.43}$$

and then to find the current density operator $\hat{\boldsymbol{J}}$ in such a way that the two together satisfy the continuity equation

$$\frac{\partial \hat{n}}{\partial t} + \nabla\cdot\hat{\boldsymbol{J}} = 0\ . \tag{2.44}$$

Using the Heisenberg equation for $\hat{n}$, which is driven by the Hamiltonian (2.28), we can rewrite the time derivative of $\hat{n}$ as

$$\frac{\partial \hat{n}}{\partial t} = -\frac{1}{i\hbar}[H, \hat{n}] \tag{2.45}$$

$$= -\frac{1}{i\hbar}\sum_\ell \frac{e_\ell}{2m_\ell}\left[\left(\boldsymbol{p}_\ell - \frac{e_\ell}{c}\boldsymbol{A}_\ell\right)^2,\ \delta(\boldsymbol{r} - \boldsymbol{r}_\ell)\right] \tag{2.46}$$

$$= -\nabla\cdot\hat{\boldsymbol{J}}\ , \tag{2.47}$$

where

$$\hat{\boldsymbol{J}} = \hat{\boldsymbol{I}} - \sum_\ell \frac{e_\ell^2}{m_\ell c}\ \boldsymbol{A}(\boldsymbol{r}_\ell, t)\ , \tag{2.48}$$

$$\hat{\boldsymbol{I}} = \sum_\ell \frac{e_\ell}{2m_\ell}\left[\boldsymbol{p}_\ell\ \delta(\boldsymbol{r} - \boldsymbol{r}_\ell) + \delta(\boldsymbol{r} - \boldsymbol{r}_\ell)\ \boldsymbol{p}_\ell\right]\ . \tag{2.49}$$

If we consider the current density due to the spin $\boldsymbol{\sigma}$ of the particle, we should add the following term [19]

$$\hat{\boldsymbol{J}}_{\rm s} = \beta\, c\, \nabla \times \sum \boldsymbol{\sigma}, \tag{2.50}$$

where $\beta\boldsymbol{\sigma}$ is the spin magnetic moment of the particle. In this case the Hamiltonian $H_{\rm M}$ should contain the corresponding Zeeman term $-\beta\,\boldsymbol{\sigma}\cdot(\nabla \times \boldsymbol{A})$ and the other relativistic correction terms.

Substituting (2.48) [plus (2.50), if necessary] into (2.42) as $\hat{b}$, we get the expectation value of the current density at $(\boldsymbol{r}, t)$ as

$$\boldsymbol{j}(\boldsymbol{r}, t) = \mathrm{Tr}\{\bar{\rho}(t)\, \hat{\boldsymbol{J}}(\boldsymbol{r}, t)\} \tag{2.51}$$

$$= \mathcal{F}[\boldsymbol{A}] \,. \tag{2.52}$$

The last equality means that $\boldsymbol{j}(\boldsymbol{r}, t)$ is a functional of $\boldsymbol{A}(\boldsymbol{r}, t)$. It should be stressed that $\boldsymbol{A}(\boldsymbol{r}, t)$ is considered as a given quantity when calculating $\boldsymbol{j}(\boldsymbol{r}, t)$ in this section.

2.3 Self-Consistent Determination of Current Density and Vector Potential

In the last two sections, we wrote the functional relationships between the induced current density $\boldsymbol{j}(\boldsymbol{r}, t)$ and the vector potential $\boldsymbol{A}(\boldsymbol{r}, t)$ as

$$\boldsymbol{j}(\boldsymbol{r}, t) = \mathcal{F}[\boldsymbol{A}] \,, \tag{2.53}$$

$$\boldsymbol{A}(\boldsymbol{r}, t) = \boldsymbol{A}_0(\boldsymbol{r}, t) + \mathcal{G}[\boldsymbol{j}] \,. \tag{2.54}$$

The former is obtained from the solution of the Schrödinger equation for a given vector potential, and the latter from the Maxwell equations for a given current density.

In any problem of radiation–matter interaction, we have to deal with the coexisting EM field and charged particles, and their mutually dependent motions are described through (2.53) and (2.54). This means that these two equations for $\boldsymbol{A}$ and $\boldsymbol{j}$ should be solved simultaneously. The problem is formulated as a scattering process. Indeed, by specifying the initial condition, we calculate $\boldsymbol{j}$ and $\boldsymbol{A}$ at any $(\boldsymbol{r}, t)$. The initial condition for the EM field appears as the incident field $\boldsymbol{A}_0$, and that for the charged particles enters through the wave function $\psi(t_0)$. The latter condition is often modified by introducing an adiabatic switch-on of the radiation–matter interaction to avoid the spurious effects associated with sudden switching, as shown in Sect. 2.5.

The requirement of self-consistency between the motions of the EM field and charged particles is not specific to this formulation. The same problem is described in QED in terms of the Heisenberg equations of motion for dynamical variables for the photon (EM field) and matter systems, which are obviously coupled with one another. Therefore the time evolution of the dynamical variables must be determined self-consistently according to a given initial condition. This scheme is itself quite similar to (2.53) and (2.54). The only difference is that the coupling is made on the level of the c-numbers in the present case, in contrast to the coupled Heisenberg equations of the q-numbers in QED. Since the starting Hamiltonian for the total system is the same, with the vector potential either quantized or unquantized, the Heisenberg equations in QED and the semiclassical equations (2.53) and (2.54) are very similar and we can establish a definite relationship between them, as will be shown in Sect. 3.7.

As a semiclassical theory, the coupling between the motions of the EM field and charged particles can be required only via c-numbers. In this respect, it would be legitimate to use relations (2.53) and (2.54). A possible extension of (2.53) will be mentioned in Sect. 3.7 through the use of transition polarizability. Since the quantum mechanical details of the matter are included in $\boldsymbol{j}(\boldsymbol{r}, t)$ on the microscopic level, the EM field $\boldsymbol{A}(\boldsymbol{r}, t)$ derived as the solution also represents the microscopic variation in space.

The functional $\mathcal{F}[\boldsymbol{A}]$ contains $\boldsymbol{A}$ to various (up to infinite) orders as described in (2.51), while the functional $\mathcal{G}[\boldsymbol{j}]$ depends on $\boldsymbol{j}$ to first order as in (2.13). It is not easy to solve the coupled equations (2.53) and (2.54) in a general form. An appropriate approximation would be to neglect the higher order terms in (2.51) according to the order of the process we are interested in. For example, if we want to describe the linear response, second harmonic generation, and four-wave mixing, we keep the expansion of (2.51) up to first, second, and third terms, respectively. This type of treatment has quite often been made when calculating nonlinear responses. However, the nonlocal character of susceptibility functions has generally been quite poorly studied in these cases. Thus, the treatment we are proposing is a straightforward extension of the usual framework for the nonlinear response to the case of nonlocal susceptibilities.

2.4 Separable Nature of Susceptibilities in Site Representation

Even in the linear response regime, solving the coupled integral equations (2.53) and (2.54) for $\boldsymbol{A}(\boldsymbol{r}, t)$ and $\boldsymbol{j}(\boldsymbol{r}, t)$ would appear to be a complicated matter, because the susceptibility is a rather complex function of two different coordinates. The case of nonlinear response should be even more complicated because of the higher complexity of the nonlinear susceptibilities.

However, a careful study of the current density as a functional of $\boldsymbol{A}(\boldsymbol{r}, t)$ leads to the general conclusion that all susceptibilities can be treated as separable integral kernels [20]. This statement is valid as a good approximation for both resonant and non-resonant cases, and for both linear and nonlinear susceptibilities. Making use of this fact, we can rewrite the integro-differential equations for $\boldsymbol{j}(\boldsymbol{r}, t)$ and $\boldsymbol{A}(\boldsymbol{r}, t)$ into a set of polynomial equations, which is certainly easier to solve. The details of this argument will be given below.

The $\boldsymbol{A}$-dependence of $\boldsymbol{j}(\boldsymbol{r}, t)$ given in (2.51) arises from three origins, i.e., $\boldsymbol{p} \cdot \boldsymbol{A}$ and $\boldsymbol{A}^2$ terms in H_{int} contributing to the density matrix $\bar{\rho}(t)$ and the $\boldsymbol{A}$-linear term in $\hat{\boldsymbol{J}}$ [see (2.48)]. All of them appear as matrix elements with respect to the many-particle wave functions. Choosing the eigenfunctions of H_0 as basis, we denote their matrix elements as

$$F_{\mu\nu}(t) = \sum_\ell \left(\frac{e_\ell}{m_\ell c}\right) \langle\mu|\boldsymbol{p}_\ell\cdot\boldsymbol{A}(\boldsymbol{r}_\ell,t)|\nu\rangle \ , \tag{2.55}$$

$$B_{\mu\nu}(t) = \sum_\ell \left(\frac{e_\ell^2}{2c^2 m_\ell}\right) \langle\mu|\boldsymbol{A}(\boldsymbol{r}_\ell,t)^2|\nu\rangle \ , \tag{2.56}$$

$$\boldsymbol{C}_{\mu\nu}(t) = \sum_\ell \left(\frac{e_\ell^2}{m_\ell c}\right) \langle\mu|\boldsymbol{A}(\boldsymbol{r}_\ell,t)|\nu\rangle \ . \tag{2.57}$$

Noting that the $\boldsymbol{A}$-independent part of $\hat{\boldsymbol{J}}$ is written as $\hat{\boldsymbol{I}}$ in (2.49), we can rewrite (2.55) as

$$F_{\mu\nu}(t) = \frac{1}{c}\int \langle\mu|\hat{\boldsymbol{I}}(\boldsymbol{r})|\nu\rangle\cdot\boldsymbol{A}(\boldsymbol{r},t)\,\mathrm{d}\boldsymbol{r} \ . \tag{2.58}$$

By introducing the two operators

$$\hat{N}(\boldsymbol{r}) = \sum_\ell \frac{e_\ell^2}{m_\ell}\delta(\boldsymbol{r}-\boldsymbol{r}_\ell) \ , \tag{2.59}$$

$$\begin{aligned}\hat{\boldsymbol{R}}(\boldsymbol{r}) &= \sum_\ell e_\ell \boldsymbol{r}_\ell \delta(\boldsymbol{r}-\boldsymbol{r}_\ell) \\ &= \boldsymbol{r}\hat{n}(\boldsymbol{r}) \ ,\end{aligned} \tag{2.60}$$

we can rewrite (2.56) and (2.57) as

$$B_{\mu\nu}(t) = \frac{1}{2c^2}\int \mathrm{d}\boldsymbol{r}\,\langle\mu|\hat{N}(\boldsymbol{r})|\nu\rangle\boldsymbol{A}(\boldsymbol{r},t)^2 \ , \tag{2.61}$$

$$\boldsymbol{C}_{\mu\nu}(t) = \frac{1}{c}\int \mathrm{d}\boldsymbol{r}\langle\mu|\hat{N}(\boldsymbol{r})|\nu\rangle\boldsymbol{A}(\boldsymbol{r},t) \ . \tag{2.62}$$

Using the explicit forms for the operators, we derive the commutation relations

$$[\hat{\boldsymbol{R}}(\boldsymbol{r}), H_0] = -\mathrm{i}\hbar\boldsymbol{r}\nabla\cdot\hat{\boldsymbol{I}}(\boldsymbol{r}) \ , \tag{2.63}$$

$$[\hat{R}_\xi(\boldsymbol{r}), \hat{I}_\eta(\boldsymbol{r}')] = -\mathrm{i}\hbar r_\xi\left[\frac{\partial}{\partial r_\eta}\delta(\boldsymbol{r}-\boldsymbol{r}')\right]\hat{N}(\boldsymbol{r}') \ , \tag{2.64}$$

where ξ and η denote the components of Cartesian coordinates. We now introduce the following two operators in terms of the ω th Fourier component of $\boldsymbol{A}$:

$$\tilde{Q}(\omega) = \int \hat{\boldsymbol{R}}(\boldsymbol{r})\cdot\tilde{\boldsymbol{A}}(\boldsymbol{r},\omega)\,\mathrm{d}\boldsymbol{r} \ , \tag{2.65}$$

$$\tilde{F}(\omega) = \int \hat{\boldsymbol{I}}(\boldsymbol{r})\cdot\tilde{\boldsymbol{A}}(\boldsymbol{r},\omega)\,\mathrm{d}\boldsymbol{r} \ . \tag{2.66}$$

Then (2.63) and (2.64) can be written as

$$[\tilde{Q}, H_0] = -\mathrm{i}\hbar\int \boldsymbol{r}\cdot\tilde{\boldsymbol{A}}(\boldsymbol{r},\omega)\nabla\cdot\hat{\boldsymbol{I}}(\boldsymbol{r})\,\mathrm{d}\boldsymbol{r} \ , \tag{2.67}$$

$$\left[\tilde{Q}, \hat{I}_\eta(\boldsymbol{r}')\right] = -\mathrm{i}\hbar \int \boldsymbol{r}\cdot\tilde{\boldsymbol{A}}(\boldsymbol{r},\omega) \left[\frac{\partial}{\partial r_\eta}\delta(\boldsymbol{r}-\boldsymbol{r}')\right] \mathrm{d}\boldsymbol{r}\hat{N}(\boldsymbol{r}') \ . \tag{2.68}$$

Carrying out a partial integration in the above two integrals, we get the following factor in the integrands

$$\frac{\partial}{\partial r_\eta}(\boldsymbol{r}\cdot\boldsymbol{A}) = A_\eta + \sum_\xi r_\xi \frac{\partial A_\xi}{\partial r_\eta} \ . \tag{2.69}$$

When the spatial variation of the field is weak, this factor can be simplified to A_η. The meaning of this approximation will be discussed later in connection with the importance of the terms $B_{\mu\nu}$ and $\boldsymbol{C}_{\mu\nu}$. The commutation relations (2.67) and (2.68) then also simplify to

$$[\tilde{Q}(\omega), H_0] = \mathrm{i}\hbar\tilde{F}(\omega) \ , \tag{2.70}$$

$$[\tilde{Q}(\omega), \hat{I}_\eta(\boldsymbol{r}')] = \mathrm{i}\hbar\tilde{A}_\eta(\boldsymbol{r}',\omega)\hat{N}(\boldsymbol{r}') \ . \tag{2.71}$$

From the $(\mu\nu)$-matrix element of (2.70), we have

$$\tilde{Q}_{\mu\nu}(\omega) = -\mathrm{i}\hbar\frac{\tilde{F}_{\mu\nu}(\omega)}{E_\mu - E_\nu} \ . \tag{2.72}$$

Substituting this result into (2.71), we obtain

$$\tilde{\boldsymbol{A}}(\boldsymbol{r},\omega)N_{\mu\nu}(\boldsymbol{r}) = \sum_\tau \left[\frac{\tilde{F}_{\mu\tau}(\omega)\boldsymbol{I}_{\tau\nu}(\boldsymbol{r})}{E_\tau - E_\mu} + \frac{\boldsymbol{I}_{\mu\tau}(\boldsymbol{r})\tilde{F}_{\tau\nu}(\omega)}{E_\tau - E_\nu}\right] \ . \tag{2.73}$$

Multiplying this equation by $\boldsymbol{A}(\boldsymbol{r},\omega')$ and integrating over $\boldsymbol{r}$, we get

$$\int \mathrm{d}\boldsymbol{r}\, N_{\mu\nu}(\boldsymbol{r})\boldsymbol{A}(\boldsymbol{r},\omega)\cdot\boldsymbol{A}(\boldsymbol{r},\omega') = \sum_\tau \left[\frac{\tilde{F}_{\mu\tau}(\omega)\tilde{F}_{\tau\nu}(\omega')}{E_\tau - E_\mu} + \frac{\tilde{F}_{\mu\tau}(\omega')\tilde{F}_{\tau\nu}(\omega)}{E_\tau - E_\nu}\right] \ . \tag{2.74}$$

The left-hand side of (2.73) integrated over $\boldsymbol{r}$ is the ω th Fourier component of $\boldsymbol{C}_{\mu\nu}(\times c)$, and (2.74) is the $(\omega+\omega')$ th Fourier component of $B_{\mu\nu}(\times 2c^2)$. Therefore, all the matrix elements of (2.55)–(2.57) are expressed in terms of $\{\tilde{F}_{\mu\nu}(\omega)\}$. The current density $\boldsymbol{j}(\boldsymbol{r},t)$ is calculated by performing further time integrals of various products of such matrix elements according to (2.51), thereby producing frequency-dependent coefficients. This means that the current density $\boldsymbol{j}(\boldsymbol{r},t)$ is written as a polynomial series of $\tilde{F}$ s with various frequency-dependent coefficients and there is no other source of $\tilde{\boldsymbol{A}}$-dependence. Noting that each factor of $\tilde{F}$ contains $\tilde{\boldsymbol{A}}$ linearly in its integrand, we may rephrase this result as the separability of all susceptibilities as integral kernels.

Knowing that the current density $\boldsymbol{j}(\boldsymbol{r},t)$ is written as a polynomial series of the various $\tilde{F}$ s, we see from (2.54) that the vector potential $\boldsymbol{A}(\boldsymbol{r},t)$ is also calculated as a polynomial series of $\tilde{F}$ s. Substituting this form of $\boldsymbol{A}$ in the definition (2.58) of the $\tilde{F}$ s, we finally obtain a set of polynomial equations for the various $\tilde{F}$s. This set of equations contains source terms due to the free field $\boldsymbol{A}_0$ in (2.54). We thus determine the $\tilde{F}$s self-consistently by solving the set of polynomial equations for a given incident field $\boldsymbol{A}_0$. Once we obtain the solutions $\{\tilde{F}\}$, it is straightforward to calculate $\boldsymbol{j}(\boldsymbol{r}, t)$ from (2.53) and then $\boldsymbol{A}(\boldsymbol{r}, t)$ from (2.54). In this way we can solve for the optical response of a given system with proper consideration of nonlocality and microscopic structure of the EM field. The explicit forms of the equations will be given in later sections for linear and nonlinear responses.

Let us discuss the validity of the approximation in which we simplify (2.69) as $\boldsymbol{A}_\eta$. We call this the slowly-varying amplitude approximation on the microscopic scale. We may distinguish resonant and non-resonant cases. For resonant cases, the contributions of B and $\boldsymbol{C}$ are negligible compared with that of $\tilde{F}$. Therefore, $\boldsymbol{j}$ is dominated by the contribution of terms arising from (2.55) alone. Hence, $\boldsymbol{j}$ is a polynomial series in $\{\tilde{F}\}$. On the other hand, the contribution of B and $\boldsymbol{C}$ becomes comparable to that of $\tilde{F}$ in non-resonant cases. However, the condition of off-resonance precludes any predominant contribution from a particular quantum state and the field is determined by a more or less averaged contribution of many quantum states. Therefore, the field should have slow spatial variation. In short, it is always a good approximation to write $\boldsymbol{j}$ as a polynomial series in $\{\tilde{F}\}$.

The above separability was already noted in the early stages of nonlocal formulations, and has been used to study the linear and nonlinear responses of MS or NS systems. In such treatments, however, the validity of separable kernels was thought to be limited to resonant processes alone, to which the $\boldsymbol{p} \cdot \boldsymbol{A}$ type terms in H_{int} contribute. In the sense that the nonlocal nature of the optical response is mainly important in resonant cases, the omission of the other two contributions (Bs and $\boldsymbol{C}$s) in calculating susceptibilities is not so serious. However, from the standpoint of generalizing the theoretical framework, it is nice to know that all the susceptibilities can be treated as separable integral kernels to good approximation, for this allows us to extend the formulation to the case of non-resonant processes.

2.5 Linear Response

In this section we show the simplest version of the self-consistent determination of $\boldsymbol{j}(\boldsymbol{r}, t)$ and $\boldsymbol{A}(\boldsymbol{r}, t)$ in the case of linear response. This provides a prototype for nonlocal formulation of radiation–matter interactions by solving simultaneous polynomial equations, i.e., a set of linear equations in this case. The extension to nonlinear processes, which will be discussed later, is straightforwardly understood as the addition of nonlinear components and different frequency components to this set of linear equations. The case of linear response thus plays a key role in understanding the whole nonlocal formulation.

It was shown in the last section that the functional $\mathcal{F}[\boldsymbol{A}]$ can be expressed as a polynomial series in

$$\tilde{F}_{\mu\nu}(\omega) = \int \mathrm{d}\boldsymbol{r}\, \langle\mu|\hat{\boldsymbol{I}}(\boldsymbol{r})|\nu\rangle\cdot\tilde{\boldsymbol{A}}(\boldsymbol{r},\omega)\,, \tag{2.75}$$

where μ and ν are the quantum numbers of the unperturbed matter Hamiltonian H_0, ω is the frequency of the EM field, and the operator $\hat{\boldsymbol{I}}(\boldsymbol{r})$ is the current density defined in (2.49). This implies the separability of the integral kernels defining $\mathcal{F}[\boldsymbol{A}]$. The separable nature is exact if we consider only the resonant processes caused by the $\boldsymbol{p}\cdot\boldsymbol{A}$ term of the radiation–matter interaction, and approximately valid when we consider the two other sources of the $\boldsymbol{A}$-dependence, the $\boldsymbol{A}^2$ term in H_{int} [see (2.31)] and the $\boldsymbol{A}$-linear term in the current density operator [see (2.48)]. The latter consideration becomes necessary when we deal with non-resonant processes, where the resonant and non-resonant contributions in $\mathcal{F}[\boldsymbol{A}]$ are comparable. As discussed in the previous section, the condition for separability to hold in the non-resonant case is that the spatial variation of $\tilde{\boldsymbol{A}}(\boldsymbol{r},\omega)$ over microscopic distances should be negligible compared with the value of $\tilde{\boldsymbol{A}}$ itself. This condition is rather easily satisfied in non-resonant cases. It should, however, be kept in mind that the merit of the nonlocal response is most significantly realized in resonant conditions.

When treating the linear response, we retain only linear terms in $\tilde{\boldsymbol{A}}$ in the expansion of $\tilde{\boldsymbol{j}}(\boldsymbol{r},\omega)$. Assuming that the matter state was in the ground state in the remote past before the adiabatic switch-on of the light–matter interaction, we obtain the wave function of $\tilde{\psi}$ in (2.38) at time t as

$$\tilde{\psi}(t) = |0\rangle + \frac{1}{\hbar c}\sum_{\lambda}\sum_{\omega}|\lambda\rangle\frac{\tilde{F}_{\lambda 0}(\omega)}{\omega_{\lambda 0}-\omega-\mathrm{i}\delta}\exp\left[\mathrm{i}(\omega_{\lambda 0}-\omega-\mathrm{i}\delta)t\right] \tag{2.76}$$

to $O(A^1)$, where

$$\hbar\omega_{\lambda\mu} = E_\lambda - E_\mu = E_{\lambda\mu}\,, \tag{2.77}$$

and $\delta(=0^+)$ describes the adiabatic switch-on of H_{int}. In terms of this wave function, the induced current density is given by

$$\boldsymbol{j}(\boldsymbol{r},t) = \langle\tilde{\psi}(t)|\mathrm{e}^{\mathrm{i}H_0 t/\hbar}\left[\hat{\boldsymbol{I}}(\boldsymbol{r}) - \frac{1}{c}\hat{N}(\boldsymbol{r})\boldsymbol{A}(\boldsymbol{r},t)\right]\mathrm{e}^{-\mathrm{i}H_0 t/\hbar}|\tilde{\psi}(t)\rangle \tag{2.78a}$$

$$\begin{aligned} &= -\frac{1}{c}\langle 0|\hat{N}(\boldsymbol{r})|0\rangle\boldsymbol{A}(\boldsymbol{r},t) \\ &\quad + \frac{1}{\hbar c}\sum_{\lambda}\sum_{\omega}\left[\frac{\mathrm{e}^{-\mathrm{i}\omega t+\delta t}}{\omega_{\lambda 0}-\omega-\mathrm{i}\delta}\langle 0|\hat{\boldsymbol{I}}(\boldsymbol{r})|\lambda\rangle\tilde{F}_{\lambda 0}(\omega)\right. \\ &\qquad\left. + \frac{\mathrm{e}^{\mathrm{i}\omega t+\delta t}}{\omega_{\lambda 0}-\omega+\mathrm{i}\delta}\langle\lambda|\hat{\boldsymbol{I}}(\boldsymbol{r})|0\rangle\tilde{F}_{\lambda 0}(\omega)^*\right]\,. \end{aligned} \tag{2.78b}$$

Since $\boldsymbol{A}(\boldsymbol{r},t)$ is real and $\hat{\boldsymbol{I}}(\boldsymbol{r})$ is Hermitian, we have

$$\tilde{F}_{\lambda 0}(\omega)^* = \tilde{F}_{0\lambda}(-\omega) . \tag{2.79}$$

The induced current density consists of the non-resonant term proportional to the ground state charge density and the contribution from the excited states containing resonance effects. If we use the expression (2.73), we can combine these two terms and get a simple result for the linear current density as

$$\tilde{\boldsymbol{j}}^{(1)}(\boldsymbol{r},\omega) = \frac{1}{c}\sum_{\nu}\left[g_\nu(\omega)\tilde{F}_{\nu 0}(\omega)\boldsymbol{I}_{0\nu}(\boldsymbol{r}) + h_\nu(\omega)\tilde{F}_{0\nu}(\omega)\boldsymbol{I}_{\nu 0}(\boldsymbol{r})\right] , \tag{2.80}$$

where g_ν and h_ν contain both resonant and non-resonant terms,

$$g_\nu(\omega) = \frac{1}{E_{\nu 0} - \hbar\omega - \mathrm{i}0^+} - \frac{1}{E_{\nu 0}} , \tag{2.81}$$

$$h_\nu(\omega) = \frac{1}{E_{\nu 0} + \hbar\omega + \mathrm{i}0^+} - \frac{1}{E_{\nu 0}} . \tag{2.82}$$

The second terms on the right-hand side of g_ν and h_ν are due to the first term on the right-hand side of (2.78b).

The field emitted by the current density (2.80) in addition to the incident field is calculated from (2.12) by replacing $\boldsymbol{j}$ with (2.80), and it is obviously a linear function of $\{\tilde{F}\}$, viz.,

$$\tilde{\boldsymbol{A}}(\boldsymbol{r},\omega) = \tilde{\boldsymbol{A}}_0(\boldsymbol{r},\omega) + \frac{1}{c}\sum_{\nu}\left[g_\nu(\omega)\tilde{F}_{\nu 0}\tilde{\boldsymbol{A}}_{0\nu}(\boldsymbol{r},\omega) + h_\nu(\omega)\tilde{F}_{0\nu}\tilde{\boldsymbol{A}}_{\nu 0}(\boldsymbol{r},\omega)\right] , \tag{2.83}$$

where $\tilde{\boldsymbol{A}}_{\mu\nu}(\boldsymbol{r},\omega)$ is the vector potential of the light emitted by the current density $\boldsymbol{I}_{\mu\nu}(\boldsymbol{r})$ defined as

$$\tilde{\boldsymbol{A}}_{\mu\nu}(\boldsymbol{r},\omega) = \frac{1}{c}\int \mathrm{d}\boldsymbol{r}'\,\mathbf{G}^{(\mathrm{T})}(\boldsymbol{r}-\boldsymbol{r}',\omega)\cdot\boldsymbol{I}_{\mu\nu}(\boldsymbol{r}') \tag{2.84a}$$

$$= \frac{1}{c}\int \mathrm{d}\boldsymbol{r}'\,G_q(\boldsymbol{r}-\boldsymbol{r}')\,\boldsymbol{I}^{(\mathrm{T})}_{\mu\nu}(\boldsymbol{r}') . \tag{2.84b}$$

The Green functions $\mathbf{G}^{(\mathrm{T})}$ and G_q are defined in Sect. 2.1. The source current $\boldsymbol{I}^{(\mathrm{T})}_{\mu\nu}(\boldsymbol{r})$ in this expression is the transverse part of $\boldsymbol{I}_{\mu\nu}(\boldsymbol{r})$, defined as

$$\boldsymbol{I}^{(\mathrm{T})}_{\mu\nu}(\boldsymbol{r}) = \frac{V}{8\pi^3}\int \mathrm{d}\boldsymbol{k}\,\mathrm{e}^{\mathrm{i}\boldsymbol{k}\cdot\boldsymbol{r}}\left[\mathbf{1} - \hat{\boldsymbol{e}}_3(\boldsymbol{k})\hat{\boldsymbol{e}}_3(\boldsymbol{k})\right]\cdot\tilde{\boldsymbol{I}}_{\mu\nu}(\boldsymbol{k}) , \tag{2.85}$$

where $\tilde{\boldsymbol{I}}_{\mu\nu}(\boldsymbol{k})$ is the Fourier component of $\boldsymbol{I}_{\mu\nu}(\boldsymbol{r})$. Of course, this satisfies

$$\nabla\cdot\boldsymbol{I}^{(\mathrm{T})}_{\mu\nu}(\boldsymbol{r}) = 0 .$$

We now obtain a set of linear equations for $\{\tilde{F}\}$ by substituting (2.83) into (2.75):

$$\tilde{F}_{\mu 0}^{(0)} = \sum_{\nu} \left[(\delta_{\mu\nu} + \mathcal{A}_{\mu 0, 0\nu}\, g_\nu)\, \tilde{F}_{\nu 0} + \mathcal{A}_{\mu 0, \nu 0}\, h_\nu\, \tilde{F}_{0\nu} \right] , \tag{2.86}$$

$$\tilde{F}_{0\mu}^{(0)} = \sum_{\nu} \left[\mathcal{A}_{0\mu, 0\nu}\, g_\nu\, \tilde{F}_{\nu 0} + (\delta_{\mu\nu} + \mathcal{A}_{0\mu, \nu 0}\, h_\nu)\, \tilde{F}_{0\nu} \right] . \tag{2.87}$$

We can set the linear equations into matrix form

$$\boldsymbol{S}\tilde{\boldsymbol{F}} = \tilde{\boldsymbol{F}}^{(0)} , \tag{2.88}$$

where

$$\boldsymbol{S} = \begin{bmatrix} \delta_{\mu\nu} + \mathcal{A}_{\mu 0, 0\nu}\, g_\nu & \mathcal{A}_{\mu 0, \nu 0}\, h_\nu \\ \mathcal{A}_{0\mu, 0\nu}\, g_\nu & \delta_{\mu\nu} + \mathcal{A}_{0\mu, \nu 0}\, h_\nu \end{bmatrix} , \tag{2.89}$$

$$\tilde{\boldsymbol{F}} = \left[\ldots, \tilde{F}_{\mu 0}, \ldots, \tilde{F}_{0\mu}, \ldots \right]^{\mathrm{T}} , \tag{2.90}$$

$$\tilde{\boldsymbol{F}}^{(0)} = \left[\ldots, \tilde{F}_{\mu 0}^{(0)}, \ldots, \tilde{F}_{0\mu}^{(0)}, \ldots \right]^{\mathrm{T}} , \tag{2.91}$$

$$\tilde{F}_{\mu\nu}^{(0)}(\omega) = \int \mathrm{d}\boldsymbol{r}\, \langle \mu | \boldsymbol{I}(\boldsymbol{r}) | \nu \rangle \cdot \tilde{\boldsymbol{A}}_0(\boldsymbol{r}, \omega) . \tag{2.92}$$

Here $\mathcal{A}_{\mu\nu,\tau\sigma}$ is defined as

$$\mathcal{A}_{\mu\nu,\tau\sigma}(\omega) = -\frac{1}{c^2} \int \mathrm{d}\boldsymbol{r} \int \mathrm{d}\boldsymbol{r}'\, \langle \mu | \boldsymbol{I}(\boldsymbol{r}) | \nu \rangle \cdot \mathbf{G}^{(\mathrm{T})}(\boldsymbol{r} - \boldsymbol{r}', \omega) \cdot \langle \tau | \boldsymbol{I}(\boldsymbol{r}') | \sigma \rangle . \tag{2.93}$$

This describes the radiative correction due to the retarded interaction among the induced current density components. In other words, it is the interaction energy between the two components via the transverse EM field of frequency ω.

The solution of (2.88) is obtained in the form

$$\tilde{\boldsymbol{F}} = \boldsymbol{S}^{-1} \tilde{\boldsymbol{F}}^{(0)} , \tag{2.94}$$

where $\tilde{\boldsymbol{F}}^{(0)}$ is related to the incident field $\tilde{\boldsymbol{A}}_0$ as in (2.92). Both the current density (2.80) and vector potential (2.83) are expressed as linear combinations of this solution $\tilde{\boldsymbol{F}}$.

In this way we can determine the response field and the corresponding current density for arbitrary $(\boldsymbol{r}, \omega)$. Reflecting the microscopic treatment of the current density matrix elements (without carrying out a coarse graining procedure), the $\boldsymbol{r}$-dependences of $\tilde{\boldsymbol{j}}(\boldsymbol{r}, \omega)$ and $\tilde{\boldsymbol{A}}(\boldsymbol{r}, \omega)$ are microscopic, in contrast to the result from traditional semiclassical response theory.

As we will discuss in detail later, the matrix $\mathcal{A}_{\mu\nu,\tau\sigma}$ takes care of the radiative interaction among components of the excited current density. The latter shifts and broadens the resonances of the response spectrum. The shift and broadening are caused by virtual emission and absorption of light by excited eigenstates of matter, which is the physical picture obtained through solving (2.88) by iteration with respect to $\mathcal{A}_{\mu\nu,\tau\sigma}$. It should be noted that the radiative correction appears in the response field $\tilde{\boldsymbol{A}}(\boldsymbol{r},\omega)$ and $\tilde{\boldsymbol{j}}(\boldsymbol{r},\omega)$, and not in the susceptibility function. The linear susceptibility relates $\tilde{\boldsymbol{j}}^{(1)}$ and $\tilde{\boldsymbol{A}}$ by

$$\tilde{\boldsymbol{j}}^{(1)}(\boldsymbol{r},\omega) = \int \mathrm{d}\boldsymbol{r}' \tilde{\chi}^{(1)}(\boldsymbol{r},\boldsymbol{r}';\omega)\tilde{\boldsymbol{A}}(\boldsymbol{r}',\omega) , \tag{2.95}$$

and it has the form

$$\tilde{\chi}^{(1)}(\boldsymbol{r},\boldsymbol{r}';\omega) = \frac{1}{c}\sum_{\mu} \left[g_\mu(\omega)\boldsymbol{I}_{0\mu}(\boldsymbol{r})\boldsymbol{I}_{\mu 0}(\boldsymbol{r}') + h_\mu(\omega)\boldsymbol{I}_{0\mu}(\boldsymbol{r}')\boldsymbol{I}_{\mu 0}(\boldsymbol{r}) \right] . \tag{2.96}$$

In a simplified treatment of response, it is rather common to introduce both radiative and non-radiative widths as a finite imaginary part in g_μ and h_μ. However, this is allowed only for non-radiative width due to the 'heat bath', which is a part of the matter. If we used a susceptibility containing radiative corrections, and solved Maxwell's equations correctly, it would lead to a double counting of the radiative correction.

2.6 Nonlinear Response

In this section, we extend the formulation to the case of the nonlocal, nonlinear response. To this end, we need to consider the higher order terms with respect to $\{\tilde{F}_{\mu\nu}(\omega)\}$ in the expression for the induced current density up to the desired order of nonlinearity. Since each factor $\tilde{F}_{\mu\nu}(\omega)$ contains the vector potential to the first order, the n th order term of $\tilde{F}$ corresponds to the n th order nonlinear polarization.

The nonlocal, nonlinear susceptibility can be obtained by evaluating the higher order terms in the expansion of the matter wave function or density matrix with respect to the perturbation due to the light field. In doing so, we keep in mind the finite extent of the relevant wave functions, in order to maintain nonlocality. Here we show the expressions obtained using the density matrix, which corresponds to the case where the initial state of the matter is given as an ensemble. If we take the pure state $|0\rangle\langle 0|$ as the initial ensemble, it does of course lead to the same result as the expansion of the wave function mentioned in Sect. 2.5.

The equation of motion satisfied by the density matrix of matter is

$$\mathrm{i}\hbar\frac{\mathrm{d}\bar{\rho}}{\mathrm{d}t} = \left[H_0 + H_{\mathrm{int}}, \bar{\rho}\right] , \tag{2.97}$$

where H_0 is the matter Hamiltonian and H_{int} the radiation–matter interaction. Expanding $\bar{\rho}$ up to the required order

$$\bar{\rho} = \bar{\rho}_0 + \bar{\rho}_1 + \bar{\rho}_2 + \bar{\rho}_3 + \dots \tag{2.98}$$

and writing the equation of motion for each order of the perturbation H_{int}, we obtain

$$\mathrm{i}\hbar\frac{\mathrm{d}\bar{\rho}_0}{\mathrm{d}t} = [H_0,\ \bar{\rho}_0] , \tag{2.99}$$

$$\mathrm{i}\hbar\frac{\mathrm{d}\bar{\rho}_j}{\mathrm{d}t} = [H_0,\ \bar{\rho}_j] + [H_{\mathrm{int}}, \bar{\rho}_{j-1}] ,\ \ (j = 1, 2, \ldots) . \tag{2.100}$$

Defining g_j as

$$\bar{\rho}_j = \exp(-\mathrm{i}H_0 t/\hbar)\, g_j \exp(\mathrm{i}H_0 t/\hbar) , \tag{2.101}$$

we get a simple set of equations for the g_j s, viz.,

$$\mathrm{i}\hbar\frac{\mathrm{d}g_j}{\mathrm{d}t} = [H'(t),\ g_{j-1}] , \tag{2.102}$$

where $H'(t)$ is the interaction representation of H_{int} defined as

$$H'(t) = \exp(\mathrm{i}H_0 t/\hbar)\, H_{\mathrm{int}} \exp(-\mathrm{i}H_0 t/\hbar) . \tag{2.103}$$

This set of differential equations allows us to determine g_j step by step, starting from g_0. Assuming the initial ensemble to be stationary, we may require $[H_0, \bar{\rho}_0] = 0$, i.e., $g_0 = \bar{\rho}_0$. Using this to solve the above differential equations, we obtain $g_1, g_2, \ldots$, consecutively in the form

$$g_1(t) = \frac{-\mathrm{i}}{\hbar}\int_{-\infty}^{t} \mathrm{d}t_1 [H'(t_1), \bar{\rho}_0] , \tag{2.104}$$

$$g_2(t) = \left(\frac{-\mathrm{i}}{\hbar}\right)^2 \int_{-\infty}^{t} \mathrm{d}t_1 \int_{-\infty}^{t_1} \mathrm{d}t_2 \big[H'(t_1), [H'(t_2), \bar{\rho}_0]\big] , \tag{2.105}$$

$$g_3(t) = \left(\frac{-\mathrm{i}}{\hbar}\right)^3 \int_{-\infty}^{t} \mathrm{d}t_1 \int_{-\infty}^{t_1} \mathrm{d}t_2 \int_{-\infty}^{t_2} \mathrm{d}t_3 \Big[H'(t_1), \big[H'(t_2), [H'(t_3), \bar{\rho}_0]\big]\Big] . \tag{2.106}$$

Using this result, we can evaluate the ensemble average of an arbitrary physical quantity up to a desired order. The expectation value of an operator $\hat{B}$ due to the j th order perturbation of H_{int} is given as $B^{(j)} = \mathrm{Tr}\{\bar{\rho}_j(t)\hat{B}\}$, which leads to the systematic expressions

$$B^{(1)} = \left(\frac{-\mathrm{i}}{\hbar}\right) \int_{-\infty}^{t} \mathrm{d}t_1 \langle [B(t), H'(t_1)] \rangle , \tag{2.107}$$

$$B^{(2)} = \left(\frac{-\mathrm{i}}{\hbar}\right)^2 \int_{-\infty}^{t} \mathrm{d}t_1 \int_{-\infty}^{t_1} \mathrm{d}t_2 \left\langle \big[[B(t), H'(t_1)], H'(t_2)\big] \right\rangle , \tag{2.108}$$

$$B^{(3)} = \left(\frac{-\mathrm{i}}{\hbar}\right)^3 \int_{-\infty}^{t} \mathrm{d}t_1 \int_{-\infty}^{t_1} \mathrm{d}t_2 \int_{-\infty}^{t_2} \mathrm{d}t_3 \tag{2.109}$$

$$\left\langle \Big[\big[[B(t), H'(t_1)], H'(t_2)\big], H'(t_3)\Big] \right\rangle ,$$

where $B(t)$ is the interaction representation of the operator $\hat{B}$ defined similarly to $H'(t)$, and $\langle \cdots \rangle$ indicates an ensemble average $\mathrm{Tr}\{\bar{\rho}_0 \ldots\}$.

Let us consider nonlinear (e.g., third order) optical susceptibilities on the basis of the general expression given above. For that purpose, we need only replace $\hat{B}$ by the current density $\hat{\boldsymbol{J}}(\boldsymbol{r})$. As is easily seen, the expressions for the nonlinear susceptibilities are lengthy because of the multiple commutators in the multiple integrals. For this reason, and for later use, let us restrict to cases of resonant nonlinear processes, where we may neglect the A^2 term in H_{int} and $\boldsymbol{A}$-linear term in $\hat{\boldsymbol{J}}(\boldsymbol{r})$. Then the only source of $\boldsymbol{A}$-dependence is the A-linear term in H_{int}, which can be rewritten as

$$H_{\mathrm{int}} = -\frac{1}{c}\int \mathrm{d}\boldsymbol{r}\hat{\boldsymbol{I}}(\boldsymbol{r})\cdot\boldsymbol{A}(\boldsymbol{r},t) , \tag{2.110}$$

where $\hat{\boldsymbol{I}}(\boldsymbol{r})$ is the $\boldsymbol{A}$-independent part of the current density operator $\hat{\boldsymbol{J}}(\boldsymbol{r})$, defined in (2.49). Defining the Fourier transform and its inverse as

$$f(t) = \int \mathrm{d}\omega\ \tilde{f}(\omega)\mathrm{e}^{-\mathrm{i}\omega t} , \tag{2.111}$$

$$\tilde{f}(\omega) = \frac{1}{2\pi}\int \mathrm{d}t f(t)\mathrm{e}^{\mathrm{i}\omega t} , \tag{2.112}$$

the matrix elements of $H'(t)$ with respect to the eigenstates of H_0 can be written as

$$\langle\mu|H'(t)|\nu\rangle = -\frac{1}{c}\int \mathrm{d}\omega\ \tilde{F}_{\mu\nu}(\omega)\exp\left[\mathrm{i}(\omega_{\mu\nu} - \omega - \mathrm{i}\delta)t\right] , \tag{2.113}$$

where $\omega_{\mu\nu} = (E_\mu - E_\nu)/\hbar$, and $\tilde{F}_{\mu\nu}(\omega)$ is defined by (2.75).

Substituting $\hat{B} = \hat{\boldsymbol{I}}(\boldsymbol{r})$ into (2.109), we can write down the expression for the third order term of the induced current density, which is expressed in terms of the third order nonlinear, nonlocal susceptibility $\boldsymbol{j}^{(3)}(\boldsymbol{r},t)$ as

$$\begin{aligned} j^{(3)}_{\bar{\xi}}(\boldsymbol{r},t) = &\int \mathrm{d}\omega_1 \int \mathrm{d}\omega_2 \int \mathrm{d}\omega_3\ \exp\left[-\mathrm{i}(\omega_1+\omega_2+\omega_3)t\right] \\ &\int \mathrm{d}\boldsymbol{r}_1 \int \mathrm{d}\boldsymbol{r}_2 \int \mathrm{d}\boldsymbol{r}_3 \sum_\xi \sum_\eta \sum_\zeta \\ &\chi^{(3)}(\boldsymbol{r},\boldsymbol{r}_1,\boldsymbol{r}_2,\boldsymbol{r}_3;\ \omega_1,\omega_2,\omega_3)_{\bar{\xi}\xi\eta\zeta}\ A_\xi(\boldsymbol{r}_1,\omega_1)A_\eta(\boldsymbol{r}_2,\omega_2)\ A_\zeta(\boldsymbol{r}_3,\omega_3) , \end{aligned} \tag{2.114}$$

where $\bar{\xi},\xi,\eta$, and ζ each stand for one of the Cartesian coordinates x,y,z.

The detailed expressions for $\chi^{(3)}$ are obtained by integrating the sum of the eight terms (due to the expanded form of the three-fold commutator) over the time variables t_1,t_2,t_3. The three-fold commutator expands as

$$\begin{aligned} \Big[[[A,B],C],D\Big] &= \left[(AB-BA)C - C(AB-BA),D\right] \\ &= ABCD - BACD - CABD + CBAD \\ &\quad -DABC + DBAC + DCAB - DCBA . \end{aligned} \tag{2.115}$$

Defining an index s ($s = 1, \ldots, 8$) to distinguish these terms (in this order), we also divide $\boldsymbol{j}^{(3)}$ into eight terms

$$\boldsymbol{j}^{(3)} = \sum_{s=1}^{8} \boldsymbol{j}_s^{(3)} . \tag{2.116}$$

For example, the term for $s = 8$ is

$$\boldsymbol{j}_8^{(3)}(\boldsymbol{r}, t) = -\left(\frac{-\mathrm{i}}{\hbar}\right)^3 \int_{-\infty}^{t} \mathrm{d}t_1 \int_{-\infty}^{t_1} \mathrm{d}t_2 \int_{-\infty}^{t_2} \mathrm{d}t_3 \quad \langle 0|H'(t_3)H'(t_2)H'(t_1)\boldsymbol{I}(\boldsymbol{r}, t)|0\rangle . \tag{2.117}$$

From (2.113), the matrix element in the integrand is given explicitly by

$$\left(\frac{-1}{c}\right)^3 \sum_{\mu}\sum_{\nu}\sum_{\sigma} \int \mathrm{d}\omega_1 \int \mathrm{d}\omega_2 \int \mathrm{d}\omega_3 \tilde{F}_{0\mu}(\omega_3)\tilde{F}_{\mu\nu}(\omega_2)\tilde{F}_{\nu\sigma}(\omega_1) \quad \langle\sigma|\boldsymbol{I}(\boldsymbol{r})|0\rangle \mathrm{e}^{\mathrm{i}(\omega_{0\mu}-\omega_3-\mathrm{i}\delta)t_3}\, \mathrm{e}^{\mathrm{i}(\omega_{\mu\nu}-\omega_2-\mathrm{i}\delta)t_2}\, \mathrm{e}^{\mathrm{i}(\omega_{\nu\sigma}-\omega_1-\mathrm{i}\delta)t_1} \mathrm{e}^{\mathrm{i}\omega_{\sigma0}t} . \tag{2.118}$$

The three-fold time integrations, taking due care over the order of integration, lead to the expression

$$\boldsymbol{j}_8^{(3)}(\boldsymbol{r}, t) = -\frac{1}{(\hbar c)^3} \sum_{\mu}\sum_{\nu}\sum_{\sigma} \int \mathrm{d}\omega_1 \int \mathrm{d}\omega_2 \int \mathrm{d}\omega_3 \quad \exp[-\mathrm{i}(\omega_1+\omega_2+\omega_3)t] \frac{\tilde{F}_{0\mu}(\omega_3)\tilde{F}_{\mu\nu}(\omega_2)\tilde{F}_{\nu\sigma}(\omega_1)\langle\sigma|\boldsymbol{I}(\boldsymbol{r})|0\rangle}{(\omega_{0\mu}-\Omega_3)(\omega_{0\nu}-\Omega_2)(\omega_{0\sigma}-\Omega_1)} , \tag{2.119}$$

where we have used the abbreviations

$$\Omega_3 = \omega_3 + \mathrm{i}\delta , \quad \Omega_2 = \omega_3 + \omega_2 + 2\mathrm{i}\delta , \quad \Omega_1 = \omega_3 + \omega_2 + \omega_1 + 3\mathrm{i}\delta . \tag{2.120}$$

Including all the other components ($s = 1, \ldots, 7$), we obtain the complete form of $\boldsymbol{j}^{(3)}(\boldsymbol{r}, \mathrm{t})$, whose integrand in the above expression is given as the product of $\exp[-\mathrm{i}(\omega_1+\omega_2+\omega_3)t]$ and

$$\begin{aligned}
&\frac{\langle 0|\boldsymbol{I}(\boldsymbol{r})|\mu\rangle F_{\mu\nu}(\omega_1)F_{\nu\sigma}(\omega_2)F_{\sigma0}(\omega_3)}{(\omega_{\sigma0}-\Omega_3)(\omega_{\nu0}-\Omega_2)(\omega_{\mu0}-\Omega_1)} - \frac{\langle\mu|\boldsymbol{I}(\boldsymbol{r})|\nu\rangle F_{0\mu}(\omega_1)F_{\nu\sigma}(\omega_2)F_{\sigma0}(\omega_3)}{(\omega_{\sigma0}-\Omega_3)(\omega_{\nu0}-\Omega_2)(\omega_{\nu\mu}-\Omega_1)} \\
&- \frac{\langle\mu|\boldsymbol{I}(\boldsymbol{r})|\nu\rangle F_{\nu\sigma}(\omega_1)F_{0\mu}(\omega_2)F_{\sigma0}(\omega_3)}{(\omega_{\sigma0}-\Omega_3)(\omega_{\sigma\mu}-\Omega_2)(\omega_{\nu\mu}-\Omega_1)} + \frac{\langle\nu|\boldsymbol{I}(\boldsymbol{r})|\sigma\rangle F_{\mu\nu}(\omega_1)F_{0\mu}(\omega_2)F_{\sigma0}(\omega_3)}{(\omega_{\sigma0}-\Omega_3)(\omega_{\sigma\mu}-\Omega_2)(\omega_{\sigma\nu}-\Omega_1)} \\
&- \frac{\langle\mu|\boldsymbol{I}(\boldsymbol{r})|\nu\rangle F_{\nu\sigma}(\omega_1)F_{\sigma0}(\omega_2)F_{0\mu}(\omega_3)}{(\omega_{0\mu}-\Omega_3)(\omega_{\sigma\mu}-\Omega_2)(\omega_{\nu\mu}-\Omega_1)} + \frac{\langle\nu|\boldsymbol{I}(\boldsymbol{r})|\sigma\rangle F_{\mu\nu}(\omega_1)F_{\sigma0}(\omega_2)F_{0\mu}(\omega_3)}{(\omega_{0\mu}-\Omega_3)(\omega_{\sigma\mu}-\Omega_2)(\omega_{\sigma\nu}-\Omega_1)} \\
&+ \frac{\langle\nu|\boldsymbol{I}(\boldsymbol{r})|\sigma\rangle F_{\sigma0}(\omega_1)F_{\mu\nu}(\omega_2)F_{0\mu}(\omega_3)}{(\omega_{0\mu}-\Omega_3)(\omega_{0\nu}-\Omega_2)(\omega_{\sigma\nu}-\Omega_1)} - \frac{\langle\sigma|\boldsymbol{I}(\boldsymbol{r})|0\rangle F_{\nu\sigma}(\omega_1)F_{\mu\nu}(\omega_2)F_{0\mu}(\omega_3)}{(\omega_{0\mu}-\Omega_3)(\omega_{0\nu}-\Omega_2)(\omega_{0\sigma}-\Omega_1)} .
\end{aligned} \tag{2.121}$$

Comparing this result with (2.114), we obtain $\chi^{(3)}$ as

$$
\begin{aligned}
&\chi^{(3)}(\boldsymbol{r},\boldsymbol{r}_1,\boldsymbol{r}_2,\boldsymbol{r}_3;\ \omega_1,\omega_2,\omega_3)_{\bar{\xi}\xi\eta\zeta} \\
&= -\frac{1}{(\hbar c)^3}\sum_\mu\sum_\nu\sum_\sigma \left\{ \frac{\langle 0|I_{\bar{\xi}}(\boldsymbol{r})|\mu\rangle\langle\mu|I_\xi(\boldsymbol{r}_1)|\nu\rangle\langle\nu|I_\eta(\boldsymbol{r}_2)|\sigma\rangle\langle\sigma|I_\zeta(\boldsymbol{r}_3)|0\rangle}{(\omega_{\sigma 0}-\Omega_3)(\omega_{\nu 0}-\Omega_2)(\omega_{\mu 0}-\Omega_1)} \right. \\
&\quad - \frac{\langle 0|I_\xi(\boldsymbol{r}_1)|\mu\rangle\langle\mu|I_{\bar{\xi}}(\boldsymbol{r})|\nu\rangle\langle\nu|I_\eta(\boldsymbol{r}_2)|\sigma\rangle\langle\sigma|I_\zeta(\boldsymbol{r}_3)|0\rangle}{(\omega_{\sigma 0}-\Omega_3)(\omega_{\nu 0}-\Omega_2)(\omega_{\nu\mu}-\Omega_1)} \\
&\quad - \frac{\langle 0|I_\eta(\boldsymbol{r}_2)|\mu\rangle\langle\mu|I_{\bar{\xi}}(\boldsymbol{r})|\nu\rangle\langle\nu|I_\xi(\boldsymbol{r}_1)|\sigma\rangle\langle\sigma|I_\zeta(\boldsymbol{r}_3)|0\rangle}{(\omega_{\sigma 0}-\Omega_3)(\omega_{\sigma\nu}-\Omega_2)(\omega_{\nu\mu}-\Omega_1)} \\
&\quad + \frac{\langle 0|I_\eta(\boldsymbol{r}_2)|\mu\rangle\langle\mu|I_\xi(\boldsymbol{r}_1)|\nu\rangle\langle\nu|I_{\bar{\xi}}(\boldsymbol{r})|\sigma\rangle\langle\sigma|I_\zeta(\boldsymbol{r}_3)|0\rangle}{(\omega_{\sigma 0}-\Omega_3)(\omega_{\sigma\mu}-\Omega_2)(\omega_{\sigma\nu}-\Omega_1)} \\
&\quad - \frac{\langle 0|I_\zeta(\boldsymbol{r}_3)|\mu\rangle\langle\mu|I_{\bar{\xi}}(\boldsymbol{r})|\nu\rangle\langle\nu|I_\xi(\boldsymbol{r}_1)|\sigma\rangle\langle\sigma|I_\eta(\boldsymbol{r}_2)|0\rangle}{(\omega_{0\mu}-\Omega_3)(\omega_{\sigma\mu}-\Omega_2)(\omega_{\nu\mu}-\Omega_1)} \\
&\quad + \frac{\langle 0|I_\zeta(\boldsymbol{r}_3)|\mu\rangle\langle\mu|I_\xi(\boldsymbol{r}_1)|\nu\rangle\langle\nu|I_{\bar{\xi}}(\boldsymbol{r})|\sigma\rangle\langle\sigma|I_\eta(\boldsymbol{r}_2)|0\rangle}{(\omega_{0\mu}-\Omega_3)(\omega_{\sigma\mu}-\Omega_2)(\omega_{\sigma\nu}-\Omega_1)} \\
&\quad + \frac{\langle 0|I_\zeta(\boldsymbol{r}_3)|\mu\rangle\langle\mu|I_\eta(\boldsymbol{r}_2)|\nu\rangle\langle\nu|I_{\bar{\xi}}(\boldsymbol{r})|\sigma\rangle\langle\sigma|I_\xi(\boldsymbol{r}_1)|0\rangle}{(\omega_{0\mu}-\Omega_3)(\omega_{0\nu}-\Omega_2)(\omega_{\sigma\nu}-\Omega_1)} \\
&\quad \left. - \frac{\langle 0|I_\zeta(\boldsymbol{r}_3)|\mu\rangle\langle\mu|I_\eta(\boldsymbol{r}_2)|\nu\rangle\langle\nu|I_\xi(\boldsymbol{r}_1)|\sigma\rangle\langle\sigma|I_{\bar{\xi}}(\boldsymbol{r})|0\rangle}{(\omega_{0\mu}-\Omega_3)(\omega_{0\nu}-\Omega_2)(\omega_{0\sigma}-\Omega_1)} \right\} .
\end{aligned} \tag{2.122}
$$

All of the eight terms have characteristic forms of energy denominator, i.e., they generally make different contributions to resonant processes. Nonlocality appears within the coherent extension of the relevant eigenstates, as in the linear response case. Each component of the induced current density consists of the third order term in $\tilde{F}$. Thus the induced field, due to a current density such as the source term in the Maxwell equations, is also written as a third order polynomial in $\tilde{F}$. If we substitute this expression for the induced field (vector potential) into the definition (2.75) for $\tilde{F}$, we obtain a set of simultaneous cubic equations to determine the relevant components of $\tilde{F}$. This is a straightforward extension of the scheme $\mathbf{S}\tilde{\mathbf{F}} = \tilde{\mathbf{F}}^{(0)}$ in the linear response. The main difference lies in the fact that more matter eigenstates, i.e., more variables $\{F\}$, are involved, and that various frequency components are involved, in contrast to the linear response case.

It is useful to note the following point about the general structure of the coupled cubic equations. We may put them in the form $(\mathbf{S}^{(1)} + \mathbf{S}^{(3)})\tilde{\mathbf{F}} = \tilde{\mathbf{F}}^{(0)}$, where $\mathbf{S}^{(1)}$ and $\mathbf{S}^{(3)}$ are the coefficient matrices for the linear and cubic nonlinear response, respectively. If we write the equations blockwise for each frequency component, the components connecting different blocks are all F-dependent, and $F^{(0)}$ s are nonzero only for the frequencies of the incident field. Thus, for the linear process where $\mathbf{S}$ does not depend on F, all the blocks are disconnected, and it is enough to solve the block for the incident frequency, as expected. A mixing of different blocks generally occurs for nonlinear processes.

The imaginary parts in the energy denominators of the above expressions are all due to the adiabatic switching of the interaction $H_{\rm int}$ $(\delta = 0^+)$ in the present derivation. This confers a radiative shift and width upon the matter excitation energies. In order to consider non-radiative decay effects, we may phenomenologically include finite decay times for various components of the density matrix in their equations of motion. However, in the case of nonlinear processes, we must be careful to choose a consistent set of such decay times. A proper consistency argument must be made on the basis of microscopic damping mechanisms [21].

From the explicit form of the induced current density, which differs from the result of macroscopic theory, we note the possibility of its resonant enhancement in two different ways. One is the resonance in nonlinear susceptibility, which occurs at the frequencies of matter excitation energies. The other is the resonant enhancement of the factors $\tilde{F}$ contained in the current density $\boldsymbol{j}$, which occurs at the frequencies shifted from matter excitation energies by the radiative shifts associated with each matter resonance. In view of the sensitive dependence of these resonances on the size and shape of MS or NS samples, one can control the enhancement of the nonlinear signals by choosing appropriate light frequencies, and sample size and shape. This could be a new guiding principle when seeking materials with high nonlinearity in MS and NS systems. Chapter 5 will be mostly devoted to the description of this aspect as a new element in the nonlinear response theory.

3. Some General Features of Nonlocal Response Theory

3.1 Spatial Structure of the Induced Field and its Resonant Enhancement

During the process of linear response, the frequency of the EM field is generally kept unchanged, but the wavelength takes various values, or the spatial structure varies, in each region of matter. For resonant processes, in particular, the spatial variation of the induced field can become very large compared with that of the incident field in vacuum. This is due to the requirement of self-consistency between the EM field and the induced current density (polarization). A resonant excitation confers the greatest weight upon a particular excitation of matter when forming induced polarization, and the spatial structure of this induced polarization reflects the details of the wave functions contributing to this level. Hence, a microscopic spatial variation is added to the more or less smooth one due to the non-resonant (background) polarization. This microscopic component of induced polarization is reflected in the spatial variation of the induced EM field through the requirement of self-consistency.

The expression (2.83) for the response field $\tilde{\boldsymbol{A}}(\boldsymbol{r},\omega)$ explicitly reveals the situation mentioned above. The field is given as a linear combination of $\tilde{\boldsymbol{A}}_{\mu\nu}(\boldsymbol{r},\omega)$ defined by (2.84a), which is the current density induced by the transition $\{\nu \to \mu\}$. The coefficients of the linear combination are determined by the self-consistency requirement, (2.86) and (2.87), which we want to rewrite in terms of the new variables

$$\tilde{X}_{\nu 0} = g_\nu \tilde{F}_{\nu 0} \ , \tag{3.1}$$

$$\tilde{X}_{0\nu} = h_\nu \tilde{F}_{0\nu} \ , \tag{3.2}$$

as

$$\tilde{F}^{(0)}_{\mu 0} = \sum_\nu \left[\left(\frac{1}{g_\nu} \delta_{\mu\nu} + \mathcal{A}_{\mu 0, 0\nu} \right) \tilde{X}_{\nu 0} + \mathcal{A}_{\mu 0, \nu 0}\, \tilde{X}_{0\nu} \right] \ , \tag{3.3}$$

$$\tilde{F}^{(0)}_{0\mu} = \sum_\nu \left[\mathcal{A}_{0\mu, 0\nu}\, \tilde{X}_{\nu 0} + \left(\frac{1}{h_\nu}\, \delta_{\mu\nu} + \mathcal{A}_{0\mu, \nu 0} \right) \tilde{X}_{0\nu} \right] \ . \tag{3.4}$$

In the form of a matrix equation, it can be written

$$\tilde{\mathbf{S}}_{\mathrm{x}} \tilde{\boldsymbol{X}} = \tilde{\boldsymbol{F}}^{(0)} . \tag{3.5}$$

In terms of the new variables, the current density and the vector potential are given by

$$\boldsymbol{j}^{(1)}(\boldsymbol{r}, \omega) = \frac{1}{c} \sum_{\nu} \left[\tilde{X}_{\nu 0}(\omega) \boldsymbol{I}_{0\nu}(\boldsymbol{r}) + \tilde{X}_{0\nu}(\omega) \boldsymbol{I}_{\nu 0}(\boldsymbol{r}) \right] , \tag{3.6}$$

$$\tilde{\boldsymbol{A}}(\boldsymbol{r}, \omega) = \tilde{\boldsymbol{A}}_0(\boldsymbol{r}, \omega) + \frac{1}{c} \sum_{\nu} \left[\tilde{X}_{\nu 0} \tilde{\boldsymbol{A}}_{0\nu}(\boldsymbol{r}) + \tilde{X}_{0\nu} \tilde{\boldsymbol{A}}_{\nu 0}(\boldsymbol{r}) \right] , \tag{3.7}$$

where the vector potential $\tilde{\boldsymbol{A}}_{\mu\nu}(\boldsymbol{r}, \omega)$ is defined by (2.84a).

In the absence of radiative corrections, the solution of (3.5) is

$$\tilde{X}_{\nu 0} = g_{\nu} \ \tilde{F}_{\nu 0}^{(0)} , \quad \tilde{X}_{0\nu} = h_{\nu} \ \tilde{F}_{0\nu}^{(0)} . \tag{3.8}$$

Near a resonance, one of the factors $\{g_{\nu}\}$ becomes very large. Both current density and vector potential therefore receive large contributions from this term.

This scenario needs to be changed in the following manner in the presence of radiative corrections. The diagonal part of the radiative correction leads to a change in the resonance condition. Indeed, the resonance frequency is shifted from $E_{\lambda 0}$ to $E_{\lambda 0} + \mathrm{Re}[\mathcal{A}_{\lambda 0, 0\lambda}]$ and the width of the resonance changes from 0^+ to $\mathrm{Im}[\tilde{A}_{\lambda 0, 0\lambda}]$. If the magnitude of the radiative correction becomes comparable to or larger than the level spacings of the matter excitation energies, we must also consider the off-diagonal parts of the radiative correction, and this leads to various complications. First of all, the shift and broadening of the resonances should be determined by seeking the complex roots of $\det |\mathbf{S}_{\mathrm{x}}| = 0$, not by simply assigning $\mathrm{Re}[\mathcal{A}]$ s and $\mathrm{Im}[\mathcal{A}]$ s for $\{\omega = E_{\lambda 0}/\hbar\}$, respectively. In addition, the eigenstates of the whole system are no longer well described by the quantum numbers of matter excitation levels. We must now consider the field-induced mixing of matter excited states.

It is a peculiarity of the microscopic nonlocal framework that the response field of a resonant EM field contains a microscopic spatial variation, the pattern and amplitude of which specifically reflect the character of the resonance. The resonance occurs at each frequency satisfying $\det |\mathbf{S}_{\mathrm{x}}| = 0$, and this differs from a matter excitation energy by the corresponding radiative shift. If we use the macroscopic local response for the same problem, a resonance is specified only through its frequency and oscillator strength without any details of its spatial structure, so that it may attribute a completely different spatial structure to the EM field near a resonance.

The spatial structure of the response field can have a remarkable effect on nonlinear processes. In the usual theory of nonlinear processes, all matrix elements of a nonlinear process consist of electric dipole transitions under the implicit assumption of the long wavelength approximation (LWA), except for a very special case of second harmonic generation to be mentioned in the introductory part of Chap. 5. For resonant processes in MS or NS systems, there is a possibility of LWA breakdown, where a non-uniform EM field with a particular spatial pattern acquires an enhanced amplitude. In such a situation, a strong transition can be induced by the field component with this particular spatial pattern. If we use such a transition as part of a nonlinear optical process, there may be resonant enhancement of the nonlinear signal. In the local response theory, resonant enhancement of a nonlinear process is simply connected to that of nonlinear susceptibility. In the nonlocal framework, on the other hand, there is an additional mechanism of resonant enhancement, namely that of internal field amplitude mentioned above.

The enhancement of the internal field is closely connected with the radiative correction of a resonant level of matter. Therefore, it generally depends on the size, shape and internal structure of the matter. Such dependence is very sensitive, especially in MS or NS systems. This means that a resonant enhancement of nonlinear effects may arise, not only from the light frequencies, but also from a change in size, shape and internal structure of the sample. This problem will be treated in detail in Chap. 5.

3.2 Resonant Structure in the Response Spectrum: Self-Sustaining (SS) Modes

The results of Sect. 2.5 show that the response field $\tilde{\boldsymbol{A}}(\boldsymbol{r},\omega)$ in linear response is expressed as a linear combination of the components of $\{F\}$, which are the solution of (2.88), i.e.,

$$(\mathbf{F}) = (\mathbf{S})^{-1}(\mathbf{F}^{(0)}) \,. \tag{3.9}$$

Since $(\mathbf{S})^{-1}$ contains det $|\mathbf{S}|$ in its denominator, the poles of $1/\det|\mathbf{S}|$ contribute to the resonant structure of the response spectrum. The condition for determining such poles, i.e., det $|\mathbf{S}| = 0$, is also the condition for obtaining non-trivial solutions to (2.88) in the absence of an incident field ($\mathbf{F}^{(0)} = 0$). Such poles are the eigenmodes of the interacting radiation–matter system. They may be called self-sustaining (SS) modes, because they correspond to nonzero amplitudes of $\tilde{\boldsymbol{j}}(\boldsymbol{r},\omega)$ and $\tilde{\boldsymbol{A}}(\boldsymbol{r},\omega)$ supporting one another to form eigenmodes. An equivalent argument can be made in terms of the matrix $\tilde{\mathbf{S}}_{\mathrm{x}}$ defined in the last section, i.e., the condition for the SS modes is det $|\tilde{\mathbf{S}}_{\mathrm{x}}| = 0$.

Such modes can exist only for particular frequencies (with damping in general). If we neglect the radiative correction term $\mathcal{A}_{\mu\nu,\tau\sigma}$ in $\mathbf{S}$, these particular frequencies represent just the excitation energies of matter. The radiative correction term contributes, in a simple case, to the shift and broadening of the matter excitation energies, and in a more general sense, to the emergence of new coupled modes, as for example in the case of an atomic excitation coupled to a cavity mode to form a new mode with Rabi splitting.

The above argument indicates that the resonant structure of optical spectra corresponds in general to (complex) eigenfrequencies of the SS modes of the radiation–matter system. The role of the incident field is to pick up the projection of such eigenmodes according to the conditions of incidence specified by frequency, polarization, incident direction, etc., which are represented by $\{\tilde{F}_{\mu\nu}^{(0)}\}$.

This point of view is not specific to the nonlocal formulation. Similar concepts can also be established in local response theory, where the equations corresponding to (2.88) are those for the Maxwell boundary conditions. The necessary and sufficient number of such relations can be written in terms of the amplitudes $\{E_j\}$ of the electric fields (with appropriate polarization) inside and outside the matter. If one writes the terms proportional to the incident field amplitude E_0 on the right-hand side and all the rest on the left-hand side, the set of boundary conditions can be written

$$\sum_j S'_{ij} \, E_j = a_i E_0 \, , \tag{3.10}$$

where the coefficient matrix $\mathbf{S}'$ and the column vector $\boldsymbol{a}$ depend on frequency and geometrical factors. In this case, $\det|\mathbf{S}'| = 0$ determines the eigenmodes, which correspond to the resonant structure of the response spectrum described by the local susceptibility. Although (2.88) and (3.10) are physically different, their solutions for a given incident condition give the response field and induced current density (or polarization) in a mutually consistent manner, and describe the resonant structure in the response spectrum. In this sense, they play the same role in their respective frameworks. This means that $\det|\mathbf{S}'| = 0$ also defines the SS modes of the system under consideration.

Although the applicability of the local description is limited in comparison with the nonlocal one, (3.10) can be used to discuss SS modes within these limitations. This kind of interpretation was made by Fuchs et al. [22] in the context of the Fabry–Pérot interference pattern for the reflectance and transmittance spectra of a thin film of local dielectric medium. When part of the induced polarization is described by a local background susceptibility, the equations determining the response field and hence the SS modes should be modified. This will be discussed in Sect. 3.4.

We will mention various examples of SS modes below, including both local and nonlocal cases. The SS modes are the extended Lorentz oscillators in the sense mentioned in Sect. 1.5. Compared with the original Lorentz oscillators, they are not pointlike but spatially extended oscillators, and their eigenfrequencies are the matter excitation energies corrected for radiative shift and broadening. Each system of interacting radiation and matter has its own SS modes, so that it is not possible to mention the examples exhaustively. We will mention several well-known concepts established in the past that can be viewed as typical examples of SS modes. The description here will be limited to a qualitative one, but some of the topics will be treated in more detail later.

Atom in Vacuum. The simplest example of an SS mode is a light-emitting state of an excited atom. The condition $\det|\mathbf{S}| = 0$ applied to this case gives a solution for a complex frequency, the real and imaginary parts of which correspond to the excitation energy of the atom and the radiative width of the excited state, respectively. The radiative width in this formulation is the same as that from a golden rule calculation in QED [2,23]. The last statement is not restricted to this particular example, but applies to any case from the atom to a bulk crystal, as shown in Sect. 3.5.

Polariton. This is a well-known coupled mode between the EM wave and an exciton (or phonon). In an infinite 3D crystal, where the wave vector is a good quantum number, the condition $\det|\mathbf{S}| = 0$ gives polariton dispersion curves, as shown in detail later. The radiative correction (2.93) is real in this case. This real part describes the polariton dispersion curve, which exhibits a remarkable shift from that of the matter excitation (exciton or phonon) in the small wave number region. For more details, see Sect. 3.6.

Surface modes derived from the poles of the bulk dielectric function are another examples of SS modes. Surface exciton polaritons, surface phonon polaritons, and surface plasma polaritons all fall into this category.

Cavity Modes. Even when the dielectric constants of the various parts of a matter system do not have any resonances, resonant modes of the EM field can arise, influenced by the geometrical construction of the matter. Typical examples are cavity modes, waveguide modes, etc. Cavities are made of 'mirrors' confining the EM field in a specific way. This leads to a quite different spatial mode density and mode structure from those in free space. The EM field of a cavity can be determined by the boundary conditions (3.10) for the field components in different parts of the matter construction. The eigenmodes of a cavity are obtained from $\det|\mathbf{S}'| = 0$ as discussed above. The eigenmodes have complex frequencies in general, even if the dielectric constants of the cavity materials are all real quantities. The imaginary part of an eigenfrequency represents the radiative decay width of the mode. If it is small, the EM field of the mode is well confined in the cavity, while if it is large, the EM field can significantly leak out of the cavity.

The above statement applies not only to cavities but also to any form of matter. An example is the argument by Fuchs et al. [22] for a dielectric slab, which is not usually regarded as a cavity. A dielectric sphere can have many SS modes. For a radius larger than the light wavelength, these modes are well-confined in the sphere, and are called whispering gallery (WG) modes. The physical picture of the modes is the phase-matched internally multi-reflected waves. WG modes are currently arousing much interest in both fundamental research and applications [24,25].

Resonant matter in a cavity provides another interesting example of SS modes. On the one hand, this provides a test stage where the radiative lifetime of the resonant matter (an atom, for example) is modified by a change in the mode structure and amplitude of the EM field [24]. To solve $\det|\mathbf{S}| = 0$ in this case, we need to evaluate the radiative correction $\mathcal{A}$ (2.93) in terms of the radiation Green function, not in free space, but in the cavity. Once $\mathcal{A}$ is evaluated, we can obtain the change in the radiative lifetime of the atom as a function of the atom position and other geometrical factors. It certainly differs from the result for the same atom in vacuum. The radiation Green function can be obtained analytically for planar and spherical geometries [26].

On the other hand, when the matter resonance and a cavity mode frequency nearly coincide, coupled modes arise between them. This is a situation beyond the scope of the golden rule. We need to treat the particular cavity mode explicitly. If the cavity mode has no leakage, it is a model case of cavity QED, described by the coupled equations of motion for two dynamical variables describing the atom excitation and the cavity photon. Generally speaking, however, a cavity mode has a finite decay time, and sometimes several cavity modes may be involved in the description of the problem. In such a situation, it is quite useful to employ the radiation Green function of the cavity in conjunction with the nonlocal formulation of the problem, as will be discussed in detail in Sect. 4.2. Evaluating the radiative correction in terms of the radiation Green function of the cavity, we can straightforwardly obtain the coupled modes of the atomic resonance and the cavity mode. The splitting of the coupled modes is called Rabi splitting. It plays an important role in various linear and nonlinear optical processes in such a system.

X Ray in a Crystal: Dynamical Scattering Regime. X rays in crystals experience Bragg reflections. In a perfect crystal, multiple scattering processes need to be considered. The mechanism for this scattering arises from the periodic spatial structure of the dielectric constant of the crystal. In most cases, it reflects the periodic charge distribution in the ground state of the crystal. But it can also be assigned partially to the periodic structure of induced dipole moments. In the macroscopic formulation, $\det|\mathbf{S}'| = 0$ for various Fourier components of the electric field leads to X-ray eigenmodes in the dynamical scattering regime. The dispersion equation in the case of the microscopic nonlocal formulation is obtained from $\det|\mathbf{S}| = 0$, which will be discussed in Sect. 3.6.

The dispersion relation of the X ray in a crystal shows (usually very small) forbidden gaps at the Brillouin zone boundary [27]. Throughout most of the Brillouin zone, the dispersion curve is very much like the one in vacuum, but the very existence of these gaps shows that a crystal is not an empty lattice for the X ray.

The problem of an EM wave in a photonic crystal is very similar to that of an X ray in the dynamical regime. The main difference is that the coupling strength of radiation and matter is usually much stronger in photonic crystals, and this leads to a quite different dispersion curve from that in a uniform dielectric medium.

The dispersion relation for both an X ray in a typical crystal and an EM wave in a photonic crystal consists of the real part alone when the crystals are infinitely large. For finite-sized systems, and this is always the case in reality, the X ray or EM wave has a chance to escape from the crystal, causing finite lifetime broadening in the dispersion relation. An interesting example of this effect has been demonstrated in resonant nuclear Bragg scattering experiments [28] by the use of synchrotron radiation. A remarkable point is the speed-up effect of radiative decay, that is, the decay time of the resonantly excited nuclear level becomes shorter by an order of magnitude (than for an isolated nucleus) when the Bragg condition is met.

Eigenmodes with real or complex dispersion relations are all examples of SS modes, and they can be derived from the condition $\det|\mathbf{S}| = 0$ prepared appropriately for each problem. An example of resonant Bragg scattering from a finite crystal will be discussed in Sect. 4.1.4.

3.3 Generalized Radiative Correction

The framework of the nonlocal response theory described in the previous chapter is based on the Coulomb gauge. The matter Hamiltonian contains the full Coulomb interaction among the charged particles in the system, and the radiation field is purely transverse (T). All the excitations in the matter are described as oscillators with eigenfrequencies (possibly with non-radiative damping), which are the 'extended Lorentz oscillators' mentioned in the introduction. The interaction of these oscillators with the radiation field introduces the radiative correction $\mathcal{A}$, viz., (2.93), which causes shifts and broadening in their eigenfrequencies.

When considering the radiative correction, it should be noted that the interaction with the transverse EM field exists for both the T and LT mixed components of the induced current density. Its L component is equivalent to the induced charge densities through the continuity equation (2.5). The effect of the T components, which emit light, is described by the radiative correction term, and that of the L component, i.e., the induced charge densities, by the term H_C in the matter Hamiltonian, discussed in Sect. 1.4.

The term H_C can be rewritten in various forms, as already indicated in Sect. 1.4. Excitations in a matter system are accompanied by induced polarizations $\{\boldsymbol{P}_\mathrm{ind}(\boldsymbol{r})\}$, which are strictly speaking the off-diagonal matrix elements of the dipole density operator with respect to the initial and final excitation (or deexcitation) states. The energy of the matter system has a contribution from the dipole–dipole interaction among these, given by

$$H_\mathrm{dd} = -\int \mathrm{d}\boldsymbol{r} \int \mathrm{d}\boldsymbol{r}' \boldsymbol{P}_\mathrm{ind}(\boldsymbol{r}) \cdot \nabla\nabla \frac{1}{|\boldsymbol{r}-\boldsymbol{r}'|} \cdot \boldsymbol{P}'_\mathrm{ind}(\boldsymbol{r}') , \tag{3.11}$$

where the prime on one of the $\boldsymbol{P}$s indicates that $\boldsymbol{P}$ and $\boldsymbol{P}'$ are generally different. This is obtained by two partial integrations of (1.22). In terms of the induced charge densities defined by $\nabla \cdot \boldsymbol{P}_\mathrm{ind} = -\rho_\mathrm{ind}$, this interaction is equivalent to the Coulomb interaction among the induced charge densities,

$$H_{\rm cc} = \int d\boldsymbol{r} \int d\boldsymbol{r}' \frac{\rho_{\rm ind}(\boldsymbol{r})\ \rho'_{\rm ind}(\boldsymbol{r}')}{|\boldsymbol{r} - \boldsymbol{r}'|} , \tag{3.12}$$

which is one of the particular expressions for $H_{\rm C}$. For electronic excitations, it can be explicitly demonstrated that this interaction represents the e–h exchange interaction [29], where $\boldsymbol{P}_{\rm ind}$ and $\rho_{\rm ind}$ represent the matrix elements of the corresponding electronic operators with respect to the relevant states of the matter.

It is well known that this interaction $H_{\rm dd}$, or equivalently $H_{\rm cc}$ or the e–h exchange interaction, gives rise to the energy difference between the L and T modes of polarization waves [30, 31]. Among various ways of understanding the LT splitting of polarization waves, such as $H_{\rm dd}$, the e–h exchange interaction, and the interaction of the polarization with the depolarization field, the one in terms of $H_{\rm cc}$ has the widest applicability, because the induced charge density is easily defined for both electrons and lattice vibrations, and for both bulk and confined systems [14].

There are two points to be noted about $H_{\rm cc}$. In the first place, the charge density is zero for the T modes, because $\nabla\cdot\boldsymbol{P} = 0$, and they thus make no contribution to this interaction. Indeed, this interaction works only among the L and LT mixed modes. Secondly, in the case of electronic excitations (excitons), the matrix element of the charge density operator becomes zero if the excited state is a pure spin triplet, since it does not include the spin operator. Therefore, only the spin singlet components of induced charge density contribute to the e–h exchange interaction, or $H_{\rm cc}$ [13].

The radiation Green function $\mathbf{G}^{(\rm T)}$ in (2.14), appearing in the definition of $\mathcal{A}$, is transverse. Indeed, it gives the T field at point $\boldsymbol{r}$ produced by the T component of the unit current density at point $\boldsymbol{r}'$. Therefore, $\mathcal{A}$ represents the interaction energy of the T components of the induced current densities via the (T) radiation field. The interaction energy between the L components is $H_{\rm cc}$, which is included in the matter Hamiltonian.

Although this scheme is consistent by itself, we may consider another version via the unification of the two correction terms $\mathcal{A}$ and $H_{\rm cc}$. This can be done in a neat form as follows. We divide the unperturbed matter Hamiltonian into two terms,

$$H_0 = H'_0 + H_{\rm cc} . \tag{3.13}$$

Applying the rotating wave approximation to (3.5), we have a set of equations to determine $\{\tilde{X}_{\mu 0}\}$:

$$\tilde{F}^{(0)}_{\mu 0} = \sum_{\nu} \left[(E_{\nu 0} - \hbar\omega)\ \delta_{\mu\nu} + \mathcal{A}_{\mu 0, 0\nu}) \right] \tilde{X}_{\nu 0} . \tag{3.14}$$

The excitation energy $E_{\nu 0}$ is an eigenvalue of the matter Hamiltonian H_0 measured from the ground state energy. If we choose the basis functions which diagonalize H'_0, the matrix elements of $H_{\rm cc}$ are generally no longer diagonal. Then we may add this term to $\mathcal{A}$, which leads to the generalized radiative correction term defined as

$$\mathcal{A}^{(\mathrm{g})}_{\mu 0,0\nu} = \langle \mu 0 | H_{\mathrm{cc}} | 0\nu \rangle + \mathcal{A}_{\mu 0,0\nu} \; . \tag{3.15}$$

Using the continuity equation, we can rewrite $\langle \mu 0 | H_{\mathrm{cc}} | 0\nu \rangle$ in terms of the matrix elements of the current density operators as

$$\langle \mu 0 | H_{\mathrm{cc}} | 0\nu \rangle = -\frac{1}{\omega^2} \int\!\!\int \mathrm{d}\boldsymbol{r}\mathrm{d}\boldsymbol{r}' \langle \mu | \boldsymbol{I}(\boldsymbol{r}) | 0 \rangle \cdot \nabla\nabla' \frac{1}{|\boldsymbol{r}-\boldsymbol{r}'|} \cdot \langle 0 | \boldsymbol{I}(\boldsymbol{r}) | \nu \rangle \; , \tag{3.16}$$

where we have carried out two partial integrations. In this form, it is easy to add the two terms of (3.15) to yield

$$\mathcal{A}^{(\mathrm{g})}_{\mu 0,0\nu} = -\frac{1}{c^2} \int\!\!\int \mathrm{d}\boldsymbol{r}\mathrm{d}\boldsymbol{r}' \langle \mu | \boldsymbol{I}(\boldsymbol{r}) | 0 \rangle \cdot \mathbf{G}^{(\mathrm{M})}(\boldsymbol{r},\boldsymbol{r}';\omega) \cdot \langle 0 | \boldsymbol{I}(\boldsymbol{r}) | \nu \rangle \; . \tag{3.17}$$

This is almost the same expression as $\mathcal{A}_{\mu 0,0\nu}$, except that the T-type Green function $\mathbf{G}^{(\mathrm{T})}$ is replaced by

$$\mathbf{G}^{(\mathrm{M})}(\boldsymbol{r},\boldsymbol{r}';\omega) = G_q(\boldsymbol{r}-\boldsymbol{r}') \left(\mathbf{1} + \frac{1}{q^2} \nabla'\nabla' \right) \tag{3.18}$$

$$= \mathbf{G}^{(\mathrm{T})}(\boldsymbol{r},\boldsymbol{r}';\omega) + \frac{1}{q^2\,|\boldsymbol{r}-\boldsymbol{r}'|} \nabla'\nabla' \; . \tag{3.19}$$

This Green function is the solution of the Maxwell equations for a pointlike current density at $\boldsymbol{r}'$, i.e.,

$$(\nabla \times \nabla \times \; - q^2) \mathbf{G}^{(\mathrm{M})}(\boldsymbol{r},\boldsymbol{r}';\omega) = 4\pi \mathbf{1} \delta(\boldsymbol{r}-\boldsymbol{r}') \; . \tag{3.20}$$

The equations for $\{\tilde{X}_{\mu 0}\}$ are rewritten as

$$\tilde{F}^{(0)}_{\mu 0} = \sum_{\nu} \left[(E'_{\nu 0} - \hbar\omega)\, \delta_{\mu\nu} + \mathcal{A}^{(\mathrm{g})}_{\mu 0,0\nu} \right] \tilde{X}_{\nu 0} \; , \tag{3.21}$$

where $\{E'_{\nu 0}\}$ are the eigenvalues of the 'matter' Hamiltonian H'_0 measured from the ground state energy. Once we obtain the solution $\{\tilde{X}_{\nu 0}\}$, the induced current density is calculated from (3.6) and the (transverse) vector potential from (3.7). In this form it is clear that we can choose arbitrary basis functions to express the matrix elements of H'_0 and the generalized radiative correction. For any choice, we get the same SS modes and the self-consistent set of the vector potential and the induced current density.

In this way, both the instantaneous and retarded interactions among the components of the current densities are put together in the generalized radiative correction term $\mathcal{A}^{(\mathrm{g})}$. Compared with $\mathcal{A}$, the generalized $\mathcal{A}^{(\mathrm{g})}$ contains the matrix elements of H_{cc}. This additional term obviously affects only the shifts of the eigenfrequencies. The radiative width obtained from $\mathcal{A}$ and $\mathcal{A}^{(\mathrm{g})}$ is the same. In view of the fact that the Green function $\mathbf{G}^{(\mathrm{M})}$ is given for various boundary conditions in the literature [26], this version of the nonlocal formalism may sometimes be more convenient.

3.4 Background Susceptibility

In the formal description of the nonlocal response theory in Chap. 2, we have explicitly treated all components of the current density with various eigenfrequencies. These are the 'extended Lorentz oscillators' mentioned in Sect. 1.5. However, in any realistic treatment it is impossible to take all the oscillators into account explicitly, because there are infinitely many components. This is not specific to the nonlocal theory, but common to any theoretical framework dealing with resonant processes. A general recipe for this problem is to treat a small number of oscillators as dynamical variables and to regard the contribution of all the others as a background dielectric medium, which is usually described in terms of a linear susceptibility. Such a background susceptibility is treated as an adjustable parameter, and is assumed always to be local, and very often isotropic. It is a kind of 'wisdom' to avoid the formidable task of dealing explicitly with infinite degrees of freedom.

In this section, we want to discuss this feature in connection with the nonlocal response theory. When we neglect the non-resonant part of the induced polarization from explicit consideration as a dynamical variable, there must be certain compensating effects on the description of the dynamical motions of the resonant components. As examples of such effects, we will discuss:

- screening of the Coulomb interaction,
- renormalization of the radiation Green function.

The former is related to the L field, and the latter to both the L and T fields.

Before going into the details, let us recall the matter Hamiltonian and the EM field used in the nonlocal response theory as discussed in Sect. 1.4. The matter Hamiltonian is chosen to include the full Coulomb interaction among all the charged particles, and the corresponding EM field is transverse. However, the usual definition of the background susceptibility (or dielectric constant) in local form [$\sim \delta(\boldsymbol{r}-\boldsymbol{r}')$] is made with respect to the full Maxwell field $\boldsymbol{E}$ containing both transverse $\boldsymbol{E}_{\mathrm{T}}$ and longitudinal $\boldsymbol{E}_{\mathrm{L}}$ components. The nonlocal susceptibilities defined with respect to $\boldsymbol{E}$ and $\boldsymbol{E}_{\mathrm{T}}$ are generally different from each other, and their relationship can be given in a simple recurrence form, viz., (3.58) of Sect. 3.6 for crystals [32], and in a more general form for non-periodic systems [14].

3.4.1 Screening of the Coulomb Interaction

If we consider all the excited states in a formal way, we can simply write down the interaction matrix of H_{cc} for all the basis functions. However, if we divide the excited states into resonant and non-resonant components and neglect the latter, we need to consider how to deal with the off-diagonal block of the matrix between the two components. This block is nonzero for L and LT-mixed modes in general, and we cannot simply neglect it. In order to treat only the resonant components explicitly, we have to renormalize the effect of the non-resonant components in the interaction scheme of the resonant components.

The most frequently used recipe is to introduce a background dielectric with a local (isotropic) dielectric constant ϵ_b instead of the non-resonant polarizations. In this treatment, the dynamical variables contributing to the resonant ($\boldsymbol{P}_\mathrm{r}$) and non-resonant polarization ($\boldsymbol{P}_\mathrm{nr}$) are regarded as independent. Rewriting Coulomb's law $\nabla\cdot\boldsymbol{D} = 0$ in the form

$$\nabla\cdot(\boldsymbol{E} + 4\pi\boldsymbol{P}_\mathrm{nr} + 4\pi\boldsymbol{P}_\mathrm{r}) = \epsilon_\mathrm{b}\nabla\cdot\left[\boldsymbol{E} + (1/\epsilon_\mathrm{b})4\pi\boldsymbol{P}_\mathrm{r}\right] = 0 , \tag{3.22}$$

where $\boldsymbol{P}_\mathrm{nr} = \chi_\mathrm{b}\boldsymbol{E}$ and $\epsilon_\mathrm{b} = 1 + 4\pi\chi_\mathrm{b}$, we see that the (longitudinal) field produced by $\boldsymbol{P}_\mathrm{r}$ is screened by ϵ_b in the presence of the background dielectric. Therefore, H_dd (H_cc) should be multiplied by $1/\epsilon_\mathrm{b}$ when we omit the non-resonant polarizations.

The above-mentioned recipe is valid in an infinitely large system. If the system has a boundary across which ϵ_b takes different values, we need to consider also the effect of the surface charge due to the background polarization induced by the field of resonant polarizations. This is the effect of mirror charge. In considering the indirect interaction among the resonant polarizations via mirror charge effect, we need to clarify the following point.

In a simple case of mirror charge potential, the source of the mirror charge is the same charge for which we want to evaluate the potential energy. Hence, if we move the source, the point at which to evaluate the potential is also displaced. Therefore, in order to evaluate the potential energy, we need to calculate the work done in moving the charge to the position under consideration from the position of vanishing force (at infinity, for example, in the case of a semi-infinite medium). On the other hand, if we calculate the potential energy of a charge A in the field of the mirror charge of another source charge B, we have only to evaluate

$$\int \mathrm{d}\boldsymbol{r}\rho_\mathrm{A}(\boldsymbol{r})\phi_\mathrm{B}(\boldsymbol{r}) , \tag{3.23}$$

where ρ_A and ϕ_B are the charge density of A and the potential due to charge B, respectively.

The screening of H_cc due to the mirror charge induced on the background dielectric corresponds to the latter case mentioned above. This can be understood as follows. Let us consider a system of confined electrons, and assume that only the electrons contributing to certain resonances are treated as dynamical variables, with all the remaining degrees of freedom renormalized in the background polarization. The Hamiltonian of the electrons is the sum of the kinetic energies, the one-electron potential energies, and the Coulomb potential among the electrons. The interaction between the induced charge densities H_cc is derived from the Coulomb potential among the electrons, as mentioned above.

The two potential energies contain the effect of the surface charge density accompanying the background polarization induced by the distribution of the electrons. In the one-electron potential, the mirror charge effect works on the source charge itself. On the other hand, the Coulomb interaction between electrons acts on different electrons, and therefore the mirror charge of an electron acts, not on itself, but on the other electrons. Therefore, the evaluation of the mirror charge effect on $H_{\rm cc}$ should be made, not as the work from the zero force situation to the position under consideration, but in the way of expression (3.23) [33]. The 'effective' one-electron potential may contain a certain contribution from the renormalization of $H_{\rm cc}$. In that case, the evaluation of its mirror charge effect is slightly different, but the conclusion about the mirror charge effect on $H_{\rm cc}$ does not change.

The e–h attraction leading to the formation of exciton states also arises from the Coulomb interaction among individual electrons. Therefore, in a confined medium, the mirror charge effect on it, i.e., the screening due to background polarization, should be considered in the same way as the e–h exchange interaction, or $H_{\rm cc}$. Hence, the mirror charge and the electron or hole should be treated independently. When the change in $\epsilon_{\rm b}$ across the boundary is large, this effect considerably modifies the e–h attraction in the bulk, leading to a large difference in the binding energy of the e–h pair when forming an exciton state. This effect has been considered by several authors in calculating the exciton energies in confined media [34–36]. Explicit consideration of the mirror charge effect in $H_{\rm cc}$ has been made only recently by Ajiki et al. [37], although it is implicitly included in the macroscopic treatment of the Maxwell equations together with an ABC [38, 39].

3.4.2 Renormalization of the Green Function

In calculating the optical responses of various matter systems, it is rather rare to think only about resonant polarization. In most cases, we must also consider non-resonant polarization, because it affects the field inducing the resonant polarization. The presence or absence of non-resonant polarization will change the strength of the radiation–matter interaction, and also the resonant energies through the screening of $H_{\rm cc}$ mentioned in the previous section. Our purpose in this section is to describe another version of the nonlocal response theory in this situation.

This is accomplished by describing the nonresonant polarization in terms of a background (local) dielectric, and regarding this part of the matter as a medium in which to define the 'free' EM field. The remaining part of the matter consists of a small number of resonant components of polarization, which are treated as the explicit dynamical variables in the nonlocal formulation. In this version of the theory, the propagation of the 'free' EM field is described by a renormalized radiation Green function, and the radiative correction defined in terms of this Green function has an additional meaning compared with the original definition (2.93). The scenario for this approach is as follows [40].

First of all, we divide the induced polarization $\boldsymbol{P}$ into resonant ($\boldsymbol{P}_{\mathrm{r}}$) and nonresonant ($\boldsymbol{P}_{\mathrm{nr}}$) parts, and assume that the latter can be described by a constant background susceptibility χ_{b}. Then the equation determining the Maxwell electric field $\boldsymbol{E}(\boldsymbol{r},\omega)$ is

$$\nabla \times (\nabla \times \boldsymbol{E}) - q^2 \boldsymbol{E} = 4\pi q^2 (\boldsymbol{P}_{\mathrm{r}} + \boldsymbol{P}_{\mathrm{nr}}) \ . \tag{3.24}$$

The induced polarization is generally written in terms of separable integral kernels, as discussed in Sect. 2.4, and thus the susceptibility of the resonant part is also separable. In terms of the background susceptibility, the nonresonant polarization has the form

$$\boldsymbol{P}_{\mathrm{nr}}(\boldsymbol{r},\omega) = \chi_{\mathrm{b}} \Theta(\boldsymbol{r}) \boldsymbol{E}(\boldsymbol{r},\omega) \ , \tag{3.25}$$

where $\Theta(\boldsymbol{r}) = 1$ or 0 for $\boldsymbol{r}$ inside or outside the background dielectric, respectively. Note that the background susceptibility is introduced with respect to the full Maxwell field, because this is the usual way to define the background dielectric constant. A more fundamental reason for this definition is that the assumption of local character (independent of size and shape of the medium) is inappropriate for the susceptibility defined for $\boldsymbol{E}_{\mathrm{T}}$, because the matter Hamiltonian used to calculate it contains the long range interaction of induced charge densities, i.e., H_{cc}.

The above equation for $\boldsymbol{E}$ can be rewritten as

$$\nabla \times (\nabla \times \boldsymbol{E}) - q^2 \left[1 + 4\pi\chi_{\mathrm{b}} \Theta(\boldsymbol{r})\right] \boldsymbol{E} = 4\pi q^2 \boldsymbol{P}_{\mathrm{r}} \ . \tag{3.26}$$

For simple geometries of the background medium, such as a semi-infinite substrate or a single sphere, analytical solutions are known [26] for the equation of the renormalized Green function,

$$\left[\nabla \times \nabla \times -q^2 \left(1 + 4\pi\chi_{\mathrm{b}} \Theta(\boldsymbol{r})\right)\right] \mathbf{G}_{\mathrm{r}}^{(\mathrm{M})}(\boldsymbol{r},\boldsymbol{r}',\omega) = 4\pi \mathbf{1} \delta(\boldsymbol{r}-\boldsymbol{r}') \ . \tag{3.27}$$

In terms of this renormalized Green function, the solution of the above equation (3.26) for the Maxwell field is

$$\boldsymbol{E}(\boldsymbol{r}) = \boldsymbol{E}_0(\boldsymbol{r}) + q^2 \int \mathrm{d}\boldsymbol{r}' \mathbf{G}_{\mathrm{r}}^{(\mathrm{M})}(\boldsymbol{r},\boldsymbol{r}') \cdot \boldsymbol{P}_{\mathrm{r}}(\boldsymbol{r}') \ , \tag{3.28}$$

where $\boldsymbol{E}_0$ represents the solution of the homogeneous equation, i.e., the free field in the presence of the background dielectric, which may contain L as well as T components, in contrast to the case in vacuum. The T component of the field $\boldsymbol{E}_{\mathrm{T}}$ $(= \mathrm{i}q\tilde{\boldsymbol{A}})$ is obtained by subtracting from $\boldsymbol{E}$ the L component $\boldsymbol{E}_{\mathrm{L}}$, i.e., the depolarization field produced by the total polarization,

$$\boldsymbol{E}_{\mathrm{L}} = \nabla\nabla\cdot \int \mathrm{d}\boldsymbol{r}' \frac{\boldsymbol{P}_{\mathrm{r}}(\boldsymbol{r}') + \boldsymbol{P}_{\mathrm{nr}}(\boldsymbol{r}')}{|\boldsymbol{r}-\boldsymbol{r}'|} \ . \tag{3.29}$$

The result is

$$\begin{aligned} \boldsymbol{E}_{\mathrm{T}}(\boldsymbol{r}) = \boldsymbol{E}_0(\boldsymbol{r}) - \chi_{\mathrm{b}} \nabla\nabla\cdot \int \mathrm{d}\boldsymbol{r}' \frac{\boldsymbol{E}_0(\boldsymbol{r}')\Theta(\boldsymbol{r}')}{|\boldsymbol{r}-\boldsymbol{r}'|} \\ + q^2 \int \mathrm{d}\boldsymbol{r}' \ \mathbf{G}_{\mathrm{r}}^{(\mathrm{T})}(\boldsymbol{r},\boldsymbol{r}') \cdot \boldsymbol{P}_{\mathrm{r}}(\boldsymbol{r}') \ , \end{aligned} \tag{3.30}$$

where the T-type renormalized Green function is given by

$$\mathbf{G}_{\mathrm{r}}^{(\mathrm{T})}(\boldsymbol{r},\boldsymbol{r}') = \mathbf{G}_{\mathrm{r}}^{(\mathrm{M})}(\boldsymbol{r},\boldsymbol{r}') - \frac{1}{q^2}\frac{\nabla'\nabla'}{|\boldsymbol{r}-\boldsymbol{r}'|} - \chi_{\mathrm{b}}\int \mathrm{d}\boldsymbol{r}''\frac{\nabla''\nabla''}{|\boldsymbol{r}-\boldsymbol{r}''|}\cdot\Theta(\boldsymbol{r}')\mathbf{G}_{\mathrm{r}}^{(\mathrm{M})}(\boldsymbol{r}'',\boldsymbol{r}') \ . \tag{3.31}$$

It should be noted that the L field consists of three components, i.e., the L fields produced by the charge densities due to

(a) the background polarization induced by the field $\boldsymbol{E}_0$, the second term on the right-hand side of (3.30),
(b) the resonant polarization,
(c) the background polarization induced by the field produced by the resonant polarization, i.e., those terms produced by the second and third terms on the right-hand side of (3.31).

All of these L fields are included in the two terms of (3.28). The interaction between the resonant polarization and the L fields produced by itself, i.e., the L fields (b) and (c), is H_{cc}, as discussed in Sect. 3.3, and it is included in the matter Hamiltonian.

As the simplest example, let us consider the case $\Theta(\boldsymbol{r}) = 1$ everywhere. Dividing the Maxwell field into two terms

$$\boldsymbol{E} = \tilde{\boldsymbol{E}}_{\mathrm{r}} + \frac{1}{\epsilon_{\mathrm{b}}q^2}\nabla\nabla\cdot\tilde{\boldsymbol{E}}_{\mathrm{r}} \ , \tag{3.32}$$

we obtain the equation for $\tilde{\boldsymbol{E}}_{\mathrm{r}}$ as

$$(\Delta + \epsilon_{\mathrm{b}}q^2)\tilde{\boldsymbol{E}}_{\mathrm{r}} = -4\pi q^2\boldsymbol{P}_{\mathrm{r}} \ . \tag{3.33}$$

The Maxwell field is obtained in the form (3.28) with the renormalized Green function

$$\mathbf{G}_{\mathrm{r}}^{(\mathrm{M})}(\boldsymbol{r},\boldsymbol{r}') = \left(\mathbf{1} + \frac{1}{\bar{q}^2}\nabla\nabla\cdot\right)G_{\bar{q}}(\boldsymbol{r}-\boldsymbol{r}') \ , \quad (\bar{q} = q\sqrt{\epsilon_{\mathrm{b}}}) \ . \tag{3.34}$$

From this $\mathbf{G}_{\mathrm{r}}^{(\mathrm{M})}$ substituted into (3.28), we obtain

$$\nabla\cdot\boldsymbol{E}(\boldsymbol{r}) = -\frac{4\pi}{\epsilon_{\mathrm{b}}}\nabla\cdot\boldsymbol{P}(\boldsymbol{r}) \ . \tag{3.35}$$

Thus the L field included in (3.28) is given by

$$\boldsymbol{E}_{\mathrm{L}}(\boldsymbol{r}) = \frac{1}{\epsilon_{\mathrm{b}}}\nabla\nabla\cdot\int \mathrm{d}\boldsymbol{r}'\frac{\boldsymbol{P}_{\mathrm{r}}(\boldsymbol{r}')}{|\boldsymbol{r}-\boldsymbol{r}'|} = \frac{1}{\epsilon_{\mathrm{b}}}\nabla\nabla\cdot\int \mathrm{d}\boldsymbol{r}'G_0(\boldsymbol{r}-\boldsymbol{r}')\boldsymbol{P}_{\mathrm{r}}(\boldsymbol{r}') \ . \tag{3.36}$$

Hence, if we express the L field only in terms of the resonant polarization, we have the screening factor $1/\epsilon_{\mathrm{b}}$ in comparison with the case without the background dielectric. The T component of the Maxwell field also has a screening effect, because it is obtained from the Green function $\mathbf{G}_{\mathrm{r}}^{(\mathrm{M})} - \nabla\nabla\cdot G_0/\bar{q}^2$.

If we use the expression for $\boldsymbol{E}_{\rm T}$ to describe the induced current density, we can just follow the standard formulation of the nonlocal response theory [40]. However, the need to evaluate the L field in order to determine the T field as shown in (3.30) seems to be redundant, because its effect is taken into account in evaluating the matter energy $\{E_{\nu 0}\}$. A more convenient way would be to use the revised version of the nonlocal response theory discussed in Sect. 3.3. Here the generalized radiative correction takes care of both $H_{\rm cc}$ and the radiative correction due to the T field via the Green function $\mathbf{G}^{(\rm M)}$, which gives the Maxwell field. Since the renormalized Green function $\mathbf{G}_{\rm r}^{(\rm M)}$ defined above is the $\mathbf{G}^{(\rm M)}$ in Sect. 3.3 for a given boundary condition, it is suitable for the revised version of the theory.

In the presence of the background dielectric, (3.21) should be revised in the following way.

- The quantum indices μ, ν should be restricted to resonant modes.
- The effect of the background dielectric should be considered in calculating $\mathcal{A}^{(\rm g)}_{\mu 0,0\nu}$, $E'_{\nu 0}$, and $\tilde{F}^{(0)}_{\mu 0}$.

The effect on $\mathcal{A}^{(\rm g)}_{\mu 0,0\nu}$ can be treated rigorously by using $\mathbf{G}_{\rm r}^{(\rm M)}$ for $\mathbf{G}^{(\rm M)}$ when defining $\mathcal{A}^{(\rm g)}_{\mu 0,0\nu}$. This contains the contributions of $H_{\rm cc}$ in addition to the radiative correction via the purely transverse EM field. The term $H_{\rm cc}$ in the presence of the background dielectric is the interaction energy between the resonant polarization and the two L fields (b) and (c) mentioned above. Hence, it is the sum of the direct and indirect interactions between the charge densities of the resonant polarizations. The direct interaction is screened by $1/\epsilon_{\rm b}$ as in the bulk, and the indirect one by the induced surface charge density of the background dielectric.

On the other hand, the effect on $E'_{\nu 0}$ should be considered as the screening of the e–h attraction by $1/\epsilon_{\rm b}$ in the first place. In the strong confinement regime of excitons, the effect of surface charge density is also important in calculating the attractive energy of an e–h pair [34–36]. In the weak confinement regime, the effect of surface charge density on the exciton binding energy may be neglected [34], because an exciton, seen from a surface charge density, is almost a neutral particle. Finally, the effect of $\epsilon_{\rm b}$ on $\tilde{F}^{(0)}_{\mu 0}$ is calculated according to $\boldsymbol{E}_0$ in (3.28).

In this way, we have a set of equations for the resonant modes,

$$\tilde{F}^{(0)}_{\mu 0} = \sum_{\nu}{}' \left[(E'_{\nu 0} - \hbar\omega)\, \delta_{\mu\nu} + \mathcal{A}^{(\rm g)}_{\mu 0,0\nu} \right] \tilde{X}_{\nu 0} \,, \tag{3.37}$$

with the understanding that $E'_{\nu 0}$ and $\tilde{F}^{(0)}_{\mu 0}$ are appropriately calculated in the presence of the background dielectric, and

$$\mathcal{A}^{(\rm g)}_{\mu 0,0\nu} = -\frac{1}{c^2} \int\!\!\int \mathrm{d}\boldsymbol{r}\mathrm{d}\boldsymbol{r}' \langle \mu|\boldsymbol{I}(\boldsymbol{r})|0\rangle \cdot \mathbf{G}_{\rm r}^{(\rm M)}(\boldsymbol{r},\boldsymbol{r}';\omega) \cdot \langle 0|\boldsymbol{I}(\boldsymbol{r})|\nu\rangle \,. \tag{3.38}$$

The prime on the summation in (3.37) means that the sum is restricted to the resonant states.

The solution of (3.37) gives the induced current density according to (3.6), and the vector potential via (3.7). If one uses $\mathbf{G}_r^{(M)}$ instead of $\mathbf{G}^{(T)}$ in the definition (2.84a) of $\tilde{\boldsymbol{A}}_{\mu\nu}(\boldsymbol{r}, \omega)$, one obtains the full Maxwell field instead of the transverse field $\boldsymbol{E}_T (= \mathrm{i}q\tilde{\boldsymbol{A}})$. All the information about the linear response of the system with a background dielectric can be obtained from this result.

Consequently, the effect of background susceptibility can be treated neatly with the help of the renormalized radiation Green function. This scheme works well for simple geometries, where the Green function $\mathbf{G}_r^{(M)}$ is known analytically. Keller has used such Green functions in the non-retarded limit to treat the case of 'point dipoles and a semi-infinite substrate' in the context of scanning near-field optical microscopy (SNOM) [41]. The finite size of oscillators has been considered in calculating the radiative correction (retardation effect) for

- a model SNOM system, i.e., an assembly of semiconducting spheres with resonant levels on a substrate [40, 42],
- an atom-like object in a semiconducting sphere [43], which will be discussed in Sect. 4.2.

3.5 Radiative Width

In this section, we discuss the imaginary part of the radiative correction $\mathcal{A}_{\mu\nu,\tau\sigma}$ defined by (2.93). The physical meaning of this expression is the interaction energy of a current density with the transverse vector potential induced by another current density, i.e., the retarded interaction of the two current densities via the transverse EM field.

If we consider a matter excitation specified by $(\mu, 0)$, the diagonal term in the corresponding radiative correction $\mathcal{A}_{\mu 0,0\mu}$ gives the shift and broadening of the transition energy $E_\mu - E_0$ of the matter in the absence of all other transitions. Its imaginary part can be rewritten as an angular integral of the form

$$\mathrm{Im}[\mathcal{A}_{\mu 0,0\mu}] = -\frac{\omega}{4\pi c^3} \int_{k=q} \mathrm{d}\Omega_{\boldsymbol{k}} \boldsymbol{I}^{\perp}_{\mu 0}(\boldsymbol{k}) \cdot \boldsymbol{I}^{\perp}_{0\mu}(-\boldsymbol{k}) \, , \tag{3.39}$$

where $\boldsymbol{I}^{\perp}_{\mu\nu}(\boldsymbol{k})$ is the perpendicular (to $\boldsymbol{k}$) component of the $\boldsymbol{k}$ th Fourier component of $\langle\mu|\boldsymbol{I}(\boldsymbol{r})|\nu\rangle$, i.e.,

$$\boldsymbol{I}^{\perp}_{\mu\nu}(\boldsymbol{k}) = \int \mathrm{d}\boldsymbol{r}\, \mathrm{e}^{\mathrm{i}\boldsymbol{k}\cdot\boldsymbol{r}} \langle\mu|\boldsymbol{I}(\boldsymbol{r})|\nu\rangle^{(\perp)} \tag{3.40}$$

$$= \sum_{i=1}^{2} \frac{e}{m} \hat{\boldsymbol{e}}_i(\boldsymbol{k}) \langle\mu|\hat{\boldsymbol{e}}_i(\boldsymbol{k}) \cdot \sum_j \boldsymbol{p}_j \exp[\mathrm{i}\boldsymbol{k}\cdot\boldsymbol{r}_j]|\nu\rangle \, . \tag{3.41}$$

In (3.41), $\{\hat{\boldsymbol{e}}_i(\boldsymbol{k}) : i = 1, 2\}$ are two mutually orthogonal unit vectors perpendicular to $\boldsymbol{k}$. To obtain (3.39) we have used the Fourier decomposition (2.17) of $G_q(\boldsymbol{r})$ to evaluate the double integral in $\mathcal{A}$.

This result shows that the present formulation leads to the same expression for the radiative width of a two-level system in vacuum as the golden rule expression from QED [44]. Note that this holds irrespective of the long wavelength approximation (LWA). In LWA, the above expression takes the more familiar form

$$\mathrm{Im}[\mathcal{A}_{\mu 0,0\nu}] = \frac{2}{3} q^3 |M_{\mu 0}|^2 , \tag{3.42}$$

where $M_{\mu 0}$ is the transition dipole moment of this two-level system.

The radiative width is a general indicator of the coupling strength of a quantum state with the EM field, because the radiative decay rate of the state is determined by the coupling with each radiation mode. Although the oscillator strength defined as

$$f_{\mu\nu} = \frac{2}{m_0 E_{\mu\nu}} |M_{\mu\nu}|^2 \tag{3.43}$$

is more frequently used for that purpose, its usefulness is restricted to the case where LWA is a good approximation. The fact that $f_{\mu\nu}$ is defined within LWA is typically reflected in its definition. Hence, $f_{\mu\nu}$ only contains information about the matter ($\sim |M_{\mu\nu}|^2$), while $\mathcal{A}$ contains the quantities of both the matter ($\boldsymbol{j}_{\mu\nu}$) and the EM field $[\mathbf{G}^{(\mathrm{T})}(\boldsymbol{r}-\boldsymbol{r}', \omega)]$. The eigenstates of a bulk crystal, apart from localized ones due to impurities, defects, self-trapping, etc., may extend farther than the wavelengths of their resonant excitation, and this is not a situation for using LWA. Therefore, it is not appropriate to assume that the oscillator strength of an exciton is concentrated in the region near the zero wave vector. The matrix element of the radiation–matter interaction is nonzero for all wave vectors, with gradual variation of its magnitude, and the reason for the strong optical contribution from the $k \sim 0$ region is simply due to the resonance effect, i.e., the proximity of the light and exciton energies.

When LWA is valid, the matter excitation sees an essentially uniform EM field. Therefore, only those states which have nonzero oscillator strength interact with the EM field. This means that there are lots of excited matter states which do not interact with the EM field under this condition. However, when LWA becomes invalid, which is often the case in resonant processes in MS or NS systems, there are more excited states interacting with the EM field, as represented by the magnitude of $\mathrm{Im}[\mathcal{A}_{\mu 0,0\mu}]$. As shown explicitly above, the expression for $\mathrm{Im}[\mathcal{A}_{\mu 0,0\mu}]$ is proportional to $|M_{\mu\nu}|^2$ in LWA, so that it is essentially equivalent to the oscillator strength in LWA.

One might think that including the light phase factor $\exp[\mathrm{i}\boldsymbol{k} \cdot \boldsymbol{r}]$ in the definition of the dipole operator to calculate $M_{\mu\nu}$ would help to avoid the 'oscillator strength' defect beyond LWA. It would help in the presence of translational symmetry, as in a perfect crystal, but not in its absence. This is because, for resonant processes in NS matter, the $\boldsymbol{k}$ of the light field can be very different from that in vacuum, so that we do not know which value(s) of $\boldsymbol{k}$ should be used in such a recipe. An effort to make this point clearer would lead to the present formalism of nonlocal response. Thus we can regard $\mathrm{Im}[\mathcal{A}]$

as a more general indicator of the coupling strength of matter excited states, which includes the oscillator strength in the LWA limit.

It should be noted that the present formulation is not the multipole expansion of polarization (or current density), although it can describe cases beyond LWA. The spatial extension of $\boldsymbol{j}$, which may be expressed by a multipole expansion in simple cases, is treated more generally in our framework in terms of $\{\tilde{F}_{\mu\nu}\}$, and this leads to a self-consistent determination of the amplitudes and spatial structures of $\boldsymbol{j}$ and $\boldsymbol{A}$.

3.6 Radiative Shift: The Polariton

In the last section, we saw that the imaginary part of $\mathcal{A}$ includes the well known case of the golden rule expression from QED for the radiative lifetime of a two-level atom in vacuum. In this section, we discuss its counterpart $\mathrm{Re}[\mathcal{A}]$ in some detail.

$\mathrm{Re}[\mathcal{A}_{\mu 0, 0\mu}]$ describes the shift in the excitation energy $E_\mu - E_0$ of a matter system due to interaction with an EM field. As a typical example of its consequences, we will discuss the case of an exciton in an infinite 3D crystal. As we shall see below, the radiatively shifted exciton energy gives the exciton–polariton dispersion in a generalized form [32].

In an infinitely large 3D crystal, a wave vector $\boldsymbol{k}$ is always a good quantum number. All the eigenstates of the matter are associated with some wave vectors, in addition to the other quantum numbers. Any state of a crystal can be described as a linear combination of Slater determinants consisting of Bloch functions with $\boldsymbol{k}$s as quantum numbers. Reflecting the properties of Bloch functions, the matrix element of the current density $\hat{\boldsymbol{I}}$ should have the form

$$\boldsymbol{I}_{\mu\nu}(\boldsymbol{r}) = \sum_{\boldsymbol{g}} \mathrm{e}^{\mathrm{i}(\boldsymbol{k}+\boldsymbol{g})\cdot\boldsymbol{r}} \tilde{\boldsymbol{I}}_{\mu\nu}(\boldsymbol{g}) , \tag{3.44}$$

where $\boldsymbol{g}$ is a reciprocal lattice vector. The effect of lattice periodicity is reflected also in $\mathcal{A}_{\mu\nu,\tau\sigma}$, i.e., it is nonzero only when the excitations (or deexcitations) $\{\mu\nu\}$ and $\{\tau\sigma\}$ have opposite wave vectors $\{\boldsymbol{k}, -\boldsymbol{k}\}$. For such pairs of excitations, we have

$$\mathcal{A}_{\mu\nu,\tau\sigma} = \frac{-1}{c^2} \sum_{\boldsymbol{g}} \frac{4\pi V}{|\boldsymbol{k}+\boldsymbol{g}|^2 - (q + \mathrm{i}0^+)^2} \sum_{i=1}^{2} \tilde{\boldsymbol{I}}_{\mu\nu}(-\boldsymbol{g})\cdot\hat{\boldsymbol{e}}_i(\boldsymbol{k}+\boldsymbol{g}) \tilde{\boldsymbol{I}}_{\tau\sigma}(\boldsymbol{g})\cdot\hat{\boldsymbol{e}}_i(\boldsymbol{k}+\boldsymbol{g}) . \tag{3.45}$$

Here $\{\hat{\boldsymbol{e}}_i(\boldsymbol{k}+\boldsymbol{g}) : i = 1, 2\}$ are mutually orthogonal unit vectors perpendicular to $\boldsymbol{k}+\boldsymbol{g}$.

Note that $\mathrm{Im}[\mathcal{A}]$ contains $\delta(|\boldsymbol{k}+\boldsymbol{g}| - q)$ in each summand of $\boldsymbol{g}$, and these are zero for a general value of $\boldsymbol{k}$. Only when $|\boldsymbol{k}+\boldsymbol{g}| = q$ does this factor give a nonzero contribution. However, $\mathrm{Re}[\mathcal{A}]$ diverges for this condition, so that such a relationship does not satisfy the dispersion equation $\det|\mathbf{S}| = 0$, where the matrix $\mathbf{S}$ is given in (2.89). In other words, the dispersion relation is satisfied by real, finite $\{\mathcal{A}_{\mu\nu,\tau\sigma}\}$. The matrix $\mathbf{S}$ is defined with respect to the set of indices $\{\mu, \nu\}$, which represent the eigenstates of the matter, and there is a sum over the reciprocal lattice vector $\boldsymbol{g}$ in each of the matrix elements.

It is useful here to note the following identity

$$\det\left|\delta_{\mu\nu} - \sum_{n=1}^{N} a_{n\mu} c_{n\nu}\right| = \det\left|\delta_{nm} - \sum_{\mu=1}^{M} c_{n\mu} a_{m\mu}\right| , \tag{3.46}$$

where $\mathbf{a}$ and $\mathbf{c}$ are both $N \times M$ matices. The proof is given in [32].

This identity is used to rewrite the dispersion equation $\det|\mathbf{S}| = 0$ in another form, by ascribing $\{\mu, \nu\}$ to matter eigenstates and $\{n, m\}$ to $\{\boldsymbol{g}, \text{polarization index}\}$:

$$\det\Big[\Big\{|\boldsymbol{k}+\boldsymbol{g}|^2 - (q + \mathrm{i}0^+)^2\Big\}\delta_{ij}\delta_{\boldsymbol{g}\boldsymbol{g}'} \tag{3.47}$$

$$-4\pi q^2 \hat{e}_i(\boldsymbol{k}+\boldsymbol{g})\cdot\bar{\chi}_{\boldsymbol{k}}(\boldsymbol{g}, \boldsymbol{g}')\cdot\hat{e}_j(\boldsymbol{k}+\boldsymbol{g}')\Big] = 0 ,$$

where

$$\bar{\chi}_{\boldsymbol{k}}(\boldsymbol{g}, \boldsymbol{g}') = \frac{V}{\omega^2}\sum_{\nu}\Big[g_\nu \tilde{\boldsymbol{I}}_{0\nu}(\boldsymbol{g})\tilde{\boldsymbol{I}}_{\nu 0}(-\boldsymbol{g}') + h_\nu \tilde{\boldsymbol{I}}_{\nu 0}(\boldsymbol{g})\tilde{\boldsymbol{I}}_{0\nu}(-\boldsymbol{g}')\Big] . \tag{3.48}$$

In this equation $q + \mathrm{i}0^+$ can safely be written as q. This means that the radiative lifetime of polaritons in the 3D crystal is ∞. This is physically reasonable, because there is no outer space for light to escape to, that would contribute to the radiative lifetime. Thus the dispersion equation for an infinite 3D crystal is wholly determined by the real part of the radiative correction $\mathcal{A}$. This argument corroborates the one mentioned above.

The size of the matrix in the new dispersion equation is $N \times N$, where $N = 2N_g$, N_g being the number of $\boldsymbol{g}$ ($\to \infty$).

The dispersion equation obtained above can also be derived from the Maxwell equations, if we express the induced polarization in terms of the susceptibility defined with respect to the transverse part $\boldsymbol{E}_{\mathrm{T}}$ of the Maxwell field. It turns out that $\bar{\chi}$ defined in (3.48) is such a susceptibility. Because of the lattice periodicity, the polarization $\boldsymbol{P}$ and the Maxwell field (including their T and L components) can be Fourier decomposed as in (3.44). The Fourier component of the induced polarization can be described in terms of either the full Maxwell field $\tilde{\boldsymbol{E}}(\boldsymbol{k}+\boldsymbol{g}, \omega)$ or its T component $\tilde{\boldsymbol{E}}_{\mathrm{T}}(\boldsymbol{k}+\boldsymbol{g}, \omega)$ as

$$\tilde{\boldsymbol{P}}(\boldsymbol{k}+\boldsymbol{g}, \omega) = \sum_{\boldsymbol{g}'} \bar{\chi}_{\boldsymbol{k}}(\boldsymbol{g}, \boldsymbol{g}')\tilde{\boldsymbol{E}}_{\mathrm{T}}(\boldsymbol{k}+\boldsymbol{g}', \omega) , \tag{3.49}$$

$$= \sum_{\boldsymbol{g}'} \chi_{\boldsymbol{k}}(\boldsymbol{g}, \boldsymbol{g}')\tilde{\boldsymbol{E}}(\boldsymbol{k}+\boldsymbol{g}', \omega) . \tag{3.50}$$

The Fourier component of the Maxwell equations reads

$$\Big[- (\boldsymbol{k}+\boldsymbol{g}) \times (\boldsymbol{k}+\boldsymbol{g}) \times (-q^2) \Big] \tilde{\boldsymbol{E}}(\boldsymbol{k}+\boldsymbol{g},\omega) = 4\pi q^2 \tilde{\boldsymbol{P}}(\boldsymbol{k}+\boldsymbol{g},\omega) \tag{3.51}$$

$$= 4\pi q^2 \sum_{\boldsymbol{g}'} \bar{\chi}_{\boldsymbol{k}}(\boldsymbol{g},\boldsymbol{g}') \cdot \tilde{\boldsymbol{E}}_{\mathrm{T}}(\boldsymbol{k}+\boldsymbol{g}') \, . \tag{3.52}$$

Each of the Fourier components $\tilde{\boldsymbol{E}}_{\mathrm{T}}(\boldsymbol{k}+\boldsymbol{g})$ is perpendicular to $\boldsymbol{k}+\boldsymbol{g}$ and therefore consists of two components along the vectors $\hat{\mathrm{e}}_i(\boldsymbol{k}+\boldsymbol{g})$, $(i=1,2)$. Decomposing each $\tilde{\boldsymbol{E}}_{\mathrm{T}}(\boldsymbol{k}+\boldsymbol{g})$ in the above equations into the two components along $\hat{\mathrm{e}}_i(\boldsymbol{k}+\boldsymbol{g})$, $(i=1,2)$, we obtain a set of linear equations in those components. The condition for the existence of a non-trivial solution in this set of linear equations is nothing but (3.47).

Making use of the same set of equations, it is easy to get another form of the equation for the existence of a non-trivial solution, with the form of an eigenvalue equation for $1/q^2$, viz.,

$$\det\left[\left(\frac{1}{q^2} - \frac{1}{|\boldsymbol{k}+\boldsymbol{g}|^2}\right)\delta_{i,j}\delta_{\boldsymbol{g},\boldsymbol{g}'} - 4\pi \frac{\hat{e}_i(\boldsymbol{k}+\boldsymbol{g})\cdot\bar{\chi}_{\boldsymbol{k}}(\boldsymbol{g},\boldsymbol{g}')\cdot\hat{e}_j(\boldsymbol{k}+\boldsymbol{g}')}{|\boldsymbol{k}+\boldsymbol{g}||\boldsymbol{k}+\boldsymbol{g}'|}\right] = 0 \, . \tag{3.53}$$

On the other hand, if we rewrite (3.51) as

$$-(\boldsymbol{k}+\boldsymbol{g}) \times \big[(\boldsymbol{k}+\boldsymbol{g}) \times \tilde{\boldsymbol{E}}(\boldsymbol{k}+\boldsymbol{g},\omega)\big] = q^2 \tilde{\boldsymbol{D}}(\boldsymbol{k}+\boldsymbol{g},\omega) \, , \tag{3.54}$$

and eliminate $\tilde{\boldsymbol{E}}$ with the help of

$$\tilde{\boldsymbol{D}}(\boldsymbol{k}+\boldsymbol{g}) = \sum_{\boldsymbol{g}'} \big[\delta_{\boldsymbol{g}\boldsymbol{g}'} + 4\pi\chi(\boldsymbol{g},\boldsymbol{g}')\big] \tilde{\boldsymbol{E}}(\boldsymbol{k}+\boldsymbol{g}') \, , \tag{3.55}$$

we obtain another form of the dispersion equation in terms of the susceptibility $\chi_{\boldsymbol{k}}(\boldsymbol{g},\boldsymbol{g}')$ [32]:

$$0 = \det\big[|\boldsymbol{k}+\boldsymbol{g}|\hat{e}_i(\boldsymbol{k}+\boldsymbol{g})\cdot\kappa_{\boldsymbol{k}}(\boldsymbol{g},\boldsymbol{g}')\cdot\hat{e}_j(\boldsymbol{k}+\boldsymbol{g}')|\boldsymbol{k}+\boldsymbol{g}'| - q^2\delta_{ij}\delta_{\boldsymbol{g}\boldsymbol{g}'}\big] \, , \tag{3.56}$$

where $\kappa_{\boldsymbol{k}} = (\mathbf{1} + 4\pi\chi_{\boldsymbol{k}})^{-1}$. This is an eigenvalue equation for q^2.

Equations (3.53) and (3.56) are the equivalent dispersion equations for excitations in a 3D crystal with explicit consideration of the lattice periodicity. From the fact that the eigenvalues correspond to q^2 or $1/q^2$, the matrices to be diagonalized are inverse to each other. Equation (3.56) is well known, while (3.53) has been derived only recently [45, 46].

The relationship between χ and $\bar{\chi}$ is obtained from (3.49) and (3.50). The L-field produced by $\boldsymbol{P}$ is determined from $\nabla\cdot\boldsymbol{D} = 0$ as

$$\tilde{\boldsymbol{E}}_{\mathrm{L}}(\boldsymbol{k}+\boldsymbol{g}) = -4\pi \frac{(\boldsymbol{k}+\boldsymbol{g})(\boldsymbol{k}+\boldsymbol{g})}{|\boldsymbol{k}+\boldsymbol{g}|^2} \cdot \tilde{\boldsymbol{P}}(\boldsymbol{k}+\boldsymbol{g}) \, . \tag{3.57}$$

Since $\tilde{\boldsymbol{P}}$ is written in terms of $\tilde{\boldsymbol{E}}_{\mathrm{T}}$ via (3.49), we can rewrite (3.49) and (3.50) entirely in terms of $\tilde{\boldsymbol{E}}_{\mathrm{T}}$. Comparing coefficients of each Fourier component in this equation leads to

$$\bar{\chi} = \chi - 4\pi\chi\cdot\mathbf{n}\cdot\bar{\chi} \, , \tag{3.58}$$

where χ and $\bar{\chi}$ are matrices with respect to both Cartesian components and $\boldsymbol{g}$, and $\mathbf{n}$ is a diagonal matrix with respect to $(\boldsymbol{g}, \boldsymbol{g}')$ with diagonal element $(\boldsymbol{k}+\boldsymbol{g})(\boldsymbol{k}+\boldsymbol{g})/|\boldsymbol{k}+\boldsymbol{g}|^2$. The above equation is a general one for infinite crystals including nonlocality. A further generalization of this relationship is possible for non-periodic systems with nonlocality [14].

Concerning applications, the best known example is the dynamical scattering of X rays in a crystal [27]. This is generally considered under nonresonant conditions. For such a study, (3.56) has been used, and because of the weakness of the interaction, it is usually enough to use the lowest order expansion $\kappa_{\boldsymbol{k}} \sim \mathbf{1} - 4\pi\chi_{\boldsymbol{k}}$ and to consider only two $\boldsymbol{g}$s (two-wave approximation). Another example of interest is photonic bands, i.e., the dispersion of an EM field by a periodic array of dielectrics. For this purpose, both of the above equations can be used, but (3.53) provides a unique method for numerical treatment by making full use of the inverse matrix feature, which will be discussed in Sect. 4.4.

3.7 Frontier with QED: Transition Polarizability

The present framework of nonlocal response theory is based on a semiclassical treatment. Indeed, the matter system is considered quantum mechanically, while the EM field is treated as a classical quantity. The fundamental dynamical variables of this scheme are current density and vector potential. We have established two relationships between them as c-numbers, (2.53) and (2.54). These turn out to be good starting equations for describing a wide range of optical processes, i.e., both linear and nonlinear phenomena, microscopic and bulk systems, and even X-ray diffraction.

However, there is a limitation due to the very nature of the semiclassical scheme. As the c-number current density to be determined self-consistently in this scheme, we have chosen $\mathrm{Tr}\{\hat{\rho}\ \hat{j}(\boldsymbol{r})\}$, as usual in the semiclassical framework, with an initial density matrix of a canonical ensemble type or its zero temperature version $|0\rangle\langle 0|$. In this choice, because of the trace operation (Tr), the initial and final states of the matrix element are automatically the same. A linear current density of this type induced by an incident field of given frequency ω oscillates with the same frequency, and emits an EM field with the same ω. This means that Raman scattering and luminescence cannot be described in this scheme.

The difficulty mentioned above suggests another possible choice for the c-number current density. Born and Huang introduced the notion of transition polarizability to describe Raman scattering [3]. This polarizability (susceptibility) is defined with respect to the different matter states $|\mu\rangle, |\nu\rangle$ in contrast to the usual one. The induced polarization (current density) derived from this polarizability oscillates with the frequency $\omega - \omega_{\mu\nu}$, where $\hbar\omega_{\mu\nu} = E_\mu - E_\nu$.

In this way, they showed how one can describe Raman scattering in a semiclassical manner. In this sense, the transition polarizability is a link between QED and the semiclassical theory of microscopic nonlocal response.

In this section, we demonstrate the parallel structure of QED and semiclassical nonlocal theory through the very similar set of fundamental equations for the vector potential and current density. The difference is that they are the equations for operators in the former and c-numbers in the latter. If we replace all the vector potentials in the QED (operator) equations with c-number amplitudes, and take matrix elements of the remaining operator equations for $\boldsymbol{J}$, we will obtain the Born and Huang-type semiclassical equations, which allow for the transition polarizability. If we exclude the transition polarizabilities from this set of equations, we are left with the nonlocal framework of this book.

Let us consider a general coupled radiation–matter system within the nonrelativistic regime. The Hamiltonian of the total system is

$$H_{\mathrm{tot}} = H_0 + H_{\mathrm{int}} + H_{\mathrm{EM}} \,, \tag{3.59}$$

where H_0 is given by (2.30) and H_{int} by (2.31), with the understanding that $\boldsymbol{A}$ is an operator, and H_{EM} is given in quantized form as

$$H_{\mathrm{EM}} = \sum_{\boldsymbol{k}\sigma} \hbar ck(a^\dagger_{\boldsymbol{k}\sigma} a_{\boldsymbol{k}\sigma} + 1/2) \,. \tag{3.60}$$

The creation and annihilation operators of free photons $(a^\dagger, a)$ are defined as usual, with wave vector $\boldsymbol{k}$ and polarization index σ $(= 1, 2)$. Hence, the vector potential is expanded as

$$\boldsymbol{A}(\boldsymbol{r}) = \sum_{\boldsymbol{k}\sigma} \left(\frac{2\pi c\hbar}{Vk}\right)^{1/2} \hat{\boldsymbol{e}}_\sigma(\boldsymbol{k}) \mathrm{e}^{\mathrm{i}\boldsymbol{k}\cdot\boldsymbol{r}} (a_{\boldsymbol{k}\sigma} + a^\dagger_{-\boldsymbol{k}\sigma}) \,, \tag{3.61}$$

where V is a normalization volume for the EM field ($\to \infty$ at the end of calculations).

Another useful expression for H_{int} is

$$H_{\mathrm{int}} = \frac{1}{c} \int \mathrm{d}\boldsymbol{r} \hat{\boldsymbol{I}}(\boldsymbol{r}) \cdot \boldsymbol{A}(\boldsymbol{r}) + \frac{1}{2c^2} \int \mathrm{d}\boldsymbol{r} \hat{N}(\boldsymbol{r}) \, \boldsymbol{A}(\boldsymbol{r})^2 \,, \tag{3.62}$$

where $\hat{N}(\boldsymbol{r})$ is given by (2.59).

The Heisenberg equations of motion for $(a^\dagger, a)$ are

$$-\mathrm{i}\hbar \frac{d}{dt} a^\dagger_{-\boldsymbol{k}\sigma} = \hbar ck a^\dagger_{-\boldsymbol{k}\sigma} - \sqrt{\frac{2\pi\hbar}{Vck}} \int \mathrm{d}\boldsymbol{r} \boldsymbol{J}(\boldsymbol{r}) \cdot \hat{\boldsymbol{e}}_\sigma(-\boldsymbol{k}) \mathrm{e}^{-\mathrm{i}\boldsymbol{k}\cdot\boldsymbol{r}} \,, \tag{3.63}$$

$$-\mathrm{i}\hbar \frac{d}{dt} a_{\boldsymbol{k}\sigma} = -\hbar ck a_{\boldsymbol{k}\sigma} + \sqrt{\frac{2\pi\hbar}{Vck}} \int \mathrm{d}\boldsymbol{r} \boldsymbol{J}(\boldsymbol{r}) \cdot \hat{\boldsymbol{e}}_\sigma(-\boldsymbol{k}) \mathrm{e}^{-\mathrm{i}\boldsymbol{k}\cdot\boldsymbol{r}} \,, \tag{3.64}$$

where

$$\boldsymbol{J}(\boldsymbol{r}) = \boldsymbol{I}(\boldsymbol{r}) - \frac{1}{c}\hat{N}(\boldsymbol{r})\boldsymbol{A}(\boldsymbol{r}) \tag{3.65}$$

is the current density operator defined in (2.48). From the above two equations of motion, we can derive

$$\left(\frac{\mathrm{d}^2}{\mathrm{d}t^2} + c^2k^2\right)(a_{\boldsymbol{k}\sigma} + a^{\dagger}_{-\boldsymbol{k}\sigma}) = \sqrt{\frac{8\pi ck}{V\hbar}} \int \mathrm{d}\boldsymbol{r}\, \boldsymbol{J}(\boldsymbol{r})\cdot\hat{\boldsymbol{e}}_{\sigma}(-\boldsymbol{k})\mathrm{e}^{-\mathrm{i}\boldsymbol{k}\cdot\boldsymbol{r}} \,. \tag{3.66}$$

Solving this equation gives the ω th Fourier component of the vector potential as

$$\boldsymbol{A}(\boldsymbol{r},\omega) = \boldsymbol{A}_0(\boldsymbol{r},\omega) + \frac{1}{c}\int \mathrm{d}\boldsymbol{r}'\mathbf{G}_{\mathrm{T}}(\boldsymbol{r},\boldsymbol{r}';q)\cdot\hat{\boldsymbol{J}}(\boldsymbol{r}',\omega) \,, \tag{3.67}$$

where $\hat{\boldsymbol{J}}(\boldsymbol{r},\omega)$ is the ω th Fourier component of the current density operator. The dyadic Green function is defined as

$$\mathbf{G}_{\mathrm{T}}(\boldsymbol{r},\boldsymbol{r}';q) = \frac{4\pi}{V}\sum_{\boldsymbol{k}\sigma} \hat{\boldsymbol{e}}_{\sigma}(\boldsymbol{k})\frac{\mathrm{e}^{\mathrm{i}\boldsymbol{k}\cdot(\boldsymbol{r}-\boldsymbol{r}')}}{k^2-(q+\mathrm{i}\delta)^2}\hat{\boldsymbol{e}}_{\sigma}(\boldsymbol{k}) \,, \tag{3.68}$$

where $q = \omega/c$ as before, and $\delta = 0^+$ represents the adiabatic switch-on of the radiation–matter interaction. This Green function is the Fourier representation of the radiation Green function defined in Sect. 2.1.

It should be stressed that the result (3.67) is obtained without any approximation, and that it has the same form as (2.12), except that $\boldsymbol{A}$ and $\boldsymbol{J}$ are now operators.

Now we calculate the time evolution of $\boldsymbol{J}$ due to the interaction term H_{int}. The Heisenberg equation for $\boldsymbol{J}$ is

$$-\mathrm{i}\hbar\frac{\mathrm{d}\boldsymbol{J}}{\mathrm{d}t} = [H_{\mathrm{s}} + H_{\mathrm{int}}, \boldsymbol{J}] \,, \tag{3.69}$$

where $H_{\mathrm{s}} = H_0 + H_{\mathrm{EM}}$. Introducing a new variable $\boldsymbol{J}'$ by

$$\boldsymbol{J} = \exp[\mathrm{i}H_{\mathrm{s}}t/\hbar]\; \boldsymbol{J}'\; \exp[-\mathrm{i}H_{\mathrm{s}}t/\hbar] \,, \tag{3.70}$$

we can rewrite (3.69) as

$$-\mathrm{i}\hbar\frac{\mathrm{d}\boldsymbol{J}'}{\mathrm{d}t} = [H'(t), \boldsymbol{J}'] \,, \tag{3.71}$$

where

$$H'(t) = \exp[-\mathrm{i}H_{\mathrm{s}}t/\hbar]\; H_{\mathrm{int}}\; \exp[\mathrm{i}H_{\mathrm{s}}t/\hbar] \,. \tag{3.72}$$

The field dependent terms in H_{s} and H_{int} are all given through the vector potential. Therefore, iterative solution of (3.69) gives the relation between the current density and vector potential as operators.

The coupled equations (3.67) and (3.69) determine the time evolution of the operators $\boldsymbol{J}$ and $\boldsymbol{A}$. From the solution, together with the initial ensemble for the matter and field, we can determine the expectation values of various physical quantities related to the response field and induced current density.

This theoretical scheme is obviously quite similar to those we have explained as a semiclassical theory of microscopic nonlocal response. They are essentially the same in the sense that the two coupled equations for the current density and vector potential describe the response field and induced polarization under given initial conditions for the matter and field. The difference lies in whether the variables are operators or c-numbers and the importance of this difference depends on the physical quantities to be explored and their desired accuracy.

In QED, we can ask for the detailed statistical properties of the response field, i.e., the ensemble of photons in various modes, while in the semiclassical treatment the field is characterized by a certain amplitude and phase for each mode, i.e., a certain average of the statistical ensemble of photons in the mode. In this sense, the similar-looking equations for $\boldsymbol{A}$ and $\boldsymbol{J}$ in QED and the semiclassical scheme can actually be quite different. Since it is easy to describe what is done in the semiclassical scheme, i.e., a simple replacement of $\boldsymbol{A}$ and $\boldsymbol{J}$ by the corresponding c-numbers, the main question here is how this replacement can or cannot be justified by considering the approximations involved.

For this purpose, the concept of coherent state in QED [47] will be useful. For a given photon mode $(a_j^\dagger, a_j)$, a coherent state $|\alpha_j\rangle$ is an eigenstate of the annihilation operator a_j, i.e.,

$$a_j|\alpha_j\rangle = \alpha_j|\alpha_j\rangle \; . \tag{3.73}$$

Therefore, the expectation values of a_j and $a_j^\dagger$ in this state are α_j and α_j^*, respectively, and that of the vector potential of this mode is the c-number given explicitly by (3.61) by replacing a_j and $a_j^\dagger$ by α_j and α_j^*, respectively. In terms of this concept, one might describe the semiclassical approximations as follows. Instead of the detailed information about the statistical ensemble of photons in each mode, we will be concerned with the averaged amplitude and phase of the vector potential of each mode. For this purpose, we assume that each mode is averaged over a given coherent state. The averaged amplitude and phase are given via the parameters (α_j, α_j^*) of the mode, and they are treated as unknown variables to be determined from the coupled equations of the semiclassical scheme.

Although the above argument seems to work well, there is another approximation step before we arrive at the semiclassical equations for $\boldsymbol{A}$ and $\boldsymbol{J}$. Indeed, if we take the statistical average of the photon operators in (3.67) and (3.69) with respect to the coherent states of all the modes concerned, we will get the semiclassical equations as desired. This is true only for particular arrangements of a_j and $a_j^\dagger$. Because of the nonvanishing commutation relation $[a_j, a_j^\dagger] = 1$, the simple replacement of $a_j(a_j^\dagger)$ by $\alpha(\alpha^*)$ is allowed only when all the $a_j^\dagger$s in a product appear on the left-hand side of all the a_js. Since the arrangement of these operators in (3.67) and (3.69) is not always like that, a complication will arise from the commutation relation. Although any array of a_j and $a_j^\dagger$ can be put in the above-mentioned order by using the commutation relation, this rearrangement process produces many additional terms. Thus, for the semiclassical approximation, we need to neglect all the complications produced by the commutation relation.

In summary, the QED and microscopic nonlocal response schemes go hand in hand up to the fundamental equations for the vector potential and current density. The operator equations can be transformed into c-number equations via the semiclassical approximation. The meaning of the semiclassical approximation is two-fold. Firstly, we simplify the description of the photon ensemble for each mode by averaging its amplitude and phase over a coherent state. Then, in the averaging process, all the complications due to the finite commutation relation are neglected.

3.8 ABC Theory, ABC-Free Theory, and the Present Framework

As mentioned in the introduction, the development of the present framework was triggered by theoretical studies of the additional boundary condition (ABC) for excitons in bounded bulk crystals. The ABC problem was introduced long ago by Pekar [5]. The essential point of the problem is as follows. In a medium which is described in terms of a dielectric function $\epsilon(\boldsymbol{k}, \omega)$ explicitly dependent on the wave vector $\boldsymbol{k}$ as well as on the frequency ω, the dispersion equation for the transverse EM wave is $\epsilon(k, \omega) = (ck/\omega)^2$. (For simplicity, we suppose that the direction of $\boldsymbol{k}$ and the polarization are chosen in such a way that the relevant EM wave is a T mode.) If there is no k-dependence in ϵ, this equation has just one solution $\omega = ck/\sqrt{\epsilon}$ (up to a sign $\pm$). In the presence of a k-dependence, there can be two or more solutions $\omega = \omega_j(k)$ $(j = 1, 2, \ldots)$. For example, a typical case with resonance like

$$\epsilon(k, \omega) = \epsilon_b + \frac{4\pi\beta}{\omega_0 + \alpha k^2 - \omega - \mathrm{i}\gamma} \tag{3.74}$$

leads to two independent solutions $\omega = \omega_j(k)$ $(j = 1, 2)$. This means that, when a crystal is irradiated with light of frequency ω from outside (vacuum, for example), two EM waves can arise in the crystal with the same frequency but with different wave numbers.

According to the traditional response theory, the response field should be determined by requiring the Maxwell boundary conditions (MBCs) at the surface(s) for the relevant waves, i.e., the incident, reflected, transmitted waves, etc. This scheme works nicely as everybody learns in the elementary theory of Maxwell equations in the presence of a (local) dielectric medium. Indeed, the number of unknowns is the same as the number of independent MBCs, and therefore we get a unique answer.

It is obvious that the number of boundary conditions does not increase even when the number of allowed waves (with different k s) is more than two, as mentioned above. This means that the traditional approach to calculating the response field does not work. In particular, the number of unknowns is larger than the number of independent MBCs. In order to obtain a unique solution, there must be ABC(s). The problem was to find principles for determining ABC(s), and to see how the details of a sample were reflected in the form of these ABC(s).

There have been a great many studies of the ABC problem since its proposal [6–9]. There have been various types of approach based on phenomenological to microscopic viewpoints. In view of the specific nature of the problem, which involves explicit handling of a coherent quantum mechanical state (exciton), the microscopic formulation has turned out to be the best among the various approaches. An appropriate scenario for the microscopic formulation consists of the following three steps:

- One solves the quantum mechanics of a matter system with explicit boundary (slab, semi-infinite system, etc.) described in terms of a microscopic model, and obtains the energy eigenvalues and wave functions of its ground and excited states.
- One then calculates the linear susceptibility of the matter in terms of energy eigenvalues and wave functions. It depends explicitly on the positions of the source field and induced polarization, i.e., it represents a nonlocal response.
- One solves the Maxwell equations with the source polarization term described by that susceptibility.

The last step gives the EM field inside the sample together with the ABC. The EM fields inside and outside the sample are connected through the MBC. Then the field in the whole region is determined uniquely. In this procedure, it is clear that the form of the ABC is determined by the initial model for the matter.

In the early stages, several approaches performed the latter two steps starting from a simple model susceptibility [48–50]. Then treatments involving all the steps appeared with various microscopic models [51–53].

In all these treatments, the matter polarization is assumed to consist of resonant and non-resonant parts. The former is described quantum mechanically, and the latter is treated approximately via a macroscopic background dielectric constant ϵ_b. In each case, it is explicitly shown that a unique solution for the response field is obtained in terms of the MBCs and ABC(s). The need for MBCs is associated with the description of non-resonant polarization in terms of a background dielectric constant.

The next step towards generalization was to note that the Maxwell equations in the third step mentioned above can be solved without referring to an ABC [54] (ABC-free framework). This way of solving the Maxwell equations could be formally extended to a general size of slab from a semi-infinite system to a quantum well. Further generalization was possible with respect to size, shape and internal structure of a sample for the formal solution of the optical response [2]. This was the first appearance of microscopic nonlocal theory. At this stage, it was noted that no boundary condition such as MBC and ABC was needed to obtain the optical response, if one treated all polarization modes as explicit dynamical variables. The only boundary condition required for solution is the one for the direction of light propagation, which is usually chosen as an outgoing wave with respect to the source polarization, as in (2.17).

It was then noted that it was possible to include nonlinear optical processes in this nonlocal response scheme, by taking only the $\boldsymbol{A} \cdot \boldsymbol{p}$ term in the radiation–matter interaction [see (2.31)] [55]. In this case, the coupled integral equations reduce to coupled N th order polynomial equations, where N is the order of nonlinearity under consideration. This reduction is possible due to the separability of the integral kernels (nonlinear susceptibilities). A generalization of the separability argument was made by Ohfuti et al. [20]. The reduction of the fundamental equations to coupled polynomial equations often helps us to understand the structure of a problem, as in the case of linear response, and also to obtain numerical solutions.

In this way, the microscopic nonlocal response theory was developed in a general form comparable to the traditional framework of macroscopic response theory, and recognized as lying between the macroscopic theory and QED.

Different Ways of Solving Maxwell's Equations. When we divide the induced current density into resonant and non-resonant parts, and treat the latter in terms of macroscopic background susceptibility χ_{b} (or background dielectric constant ϵ_{b}), the equation for the ω th Fourier component of the Maxwell electric field is

$$\nabla \times (\nabla \times \boldsymbol{E}) - q^2[1 + 4\pi\chi_{\mathrm{b}}\, \Theta(\boldsymbol{r})]\boldsymbol{E} = 4\pi \mathrm{i}\frac{q}{c}\boldsymbol{j}_{\mathrm{r}}\,, \tag{3.75}$$

where $\Theta(\boldsymbol{r})$ is unity (zero) inside (outside) the matter, and $\boldsymbol{j}_{\mathrm{r}}$ is the resonant part of the current density.

There are evidently different ways to solve this equation, although they are essentially equivalent. The difference lies in the way χ_{b} and $\boldsymbol{j}_{\mathrm{r}}$ are treated. First, we mention the two ways of handling χ_{b}. One is the method mentioned in Sect. 3.4, where we introduce the radiation Green function (tensor) $\mathbf{G}_{\mathrm{r}}(\boldsymbol{r}, \boldsymbol{r}'; \omega)$ for the operator on the left-hand side of (3.75), i.e., (3.27). In terms of this $\mathbf{G}_{\mathrm{r}}$, the solution of (3.75) is written in a standard manner for the Green function method, as in (3.28). This is the sum of the incident field together with the field induced by $\boldsymbol{j}_{\mathrm{r}}$ and also by the non-resonant current density. To obtain an explicit form for $\mathbf{G}_{\mathrm{r}}$, the function must be smoothly connected at the boundary of the matter, which is the same condition for the EM field, i.e., MBCs.

The other way of handling χ_{b} is to solve (3.75) separately for $\Theta = 1$ and $\Theta = 0$, and connect the solutions for the two regions via MBCs. In view of the equivalence of the smooth connection of $\mathbf{G}_{\mathrm{r}}$ and the MBCs for the EM field at the matter boundary, it is obvious that the two ways of treating χ_{b}

should lead to the same answer in expressing the field induced by the current density.

The difference between the treatments of $\boldsymbol{j}_{\mathrm{r}}$ leads to different equations for determining the EM field. In particular, the nonlocal treatment leads to $\mathbf{S}\tilde{\boldsymbol{F}} = \tilde{\boldsymbol{F}}^{(0)}$ as given in (2.91), and a standard ABC theory to $\mathbf{S}'\boldsymbol{E} = E_0\boldsymbol{a}$ as given in (3.10) in Sect. 3.2. In the nonlocal theory, we make use of the fact that the resonant part of the susceptibility is described by a separable integral kernel. We evaluate the radiative correction $\mathcal{A}^{(\mathrm{g})}$ in terms of the renormalized radiation Green function and solve the coupled equations (3.21) in Sect. 3.3 for the variables $\{X_{\mu 0}\}$ defined for the resonant components alone. Although MBCs are used in obtaining the radiation Green function, the use of an ABC is not necessary, because of the nonlocal treatment of the resonant components via $\tilde{\mathbf{S}}_{\mathrm{x}}\tilde{\boldsymbol{X}} = \tilde{\boldsymbol{F}}^{(0)}$.

In many of the ABC and ABC-free theories [51, 53], the fields inside and outside the sample ($\Theta = 1$ or 0) are solved separately. This introduces free parameters originating from the homogeneous part of the differential equation in each region. Then these solutions are smoothly connected across the sample boundary, which gives a set of coupled equations for the free parameters $\mathbf{S}'\mathbf{E} = E_0\mathbf{a}$, where the components of the vectors $\mathbf{E}$ and $E^{(0)}$ represent the free parameters. The solution of these equations for a given incident light condition uniquely fixes the EM field in all the regions, and this further determines the induced current density.

Although the form of the equations in the final stages is different for the two methods mentioned above, the response should be the same, because it is derived from the same starting equation, viz., (3.75). This means that the resonant structures included in $1/\det|\mathbf{S}|$ and $1/\det|\mathbf{S}'|$ in Sect. 3.3 should be the same, i.e., both schemes should lead to the same SS (self-sustaining) modes for the coupled radiation–matter system. Furthermore, this means that the macroscopic scheme based on the matrix $\mathbf{S}'$ implicitly includes the effect of radiative shift and width, which is explicitly treated in the nonlocal scheme based on the matrix $\mathbf{S}$.

3.9 Size Enhancement of $\chi^{(3)}$. Saturation and Cancellation Problem

In this section, we discuss the size enhancement of the third-order nonlinear susceptibility within LWA. Our aim is not only to show a remarkable feature of MS or NS systems, but also to demonstrate the need for a theoretical treatment beyond LWA for understanding the intrinsic mechanism of saturation behavior that follows size enhancement. This feature of the optical response, as well as others, indicates the importance of the microscopic nonlocal response theory.

For the discussion in this section, we resume the use of electric field and polarization instead of vector potential and current density, because the main part of the argument is related to the susceptibility in LWA, or local susceptibility, which is usually treated in terms of $(\boldsymbol{E}, \boldsymbol{P})$ in macroscopic local response theory. One could of course maintain a consistent use of $(\boldsymbol{A}, \boldsymbol{j})$, but we prefer the more familiar tool in the context of this particular topic.

As for nonlinear susceptibilities, we know how they behave in two limiting cases:

(A) an assembly of independent atoms,
(B) an assembly of harmonic oscillators (free bosons).

In the former case, the susceptibilities are those of a single atom multiplied by the density of atoms, and in the latter case, all the nonlinear susceptibilities are zero. Case A is physically obvious, but the general expression for the nonlinear susceptibilities in the assembly of atoms gives many additional terms, not all of which are zero. Thus, in order to uphold our physical intuition, the additional terms must cancel each other out. It is not obvious at first glance that this is the case, but it can be shown if sufficient care is taken. Later in this section, we will give a detailed description of the cancellation problem, which commonly appears in any calculation of nonlinear susceptibility, and plays an important role in its dependence on sample size.

Case B deals with an idealized situation, where the matter Hamiltonian H_0 is described as a sum of harmonic oscillators (free bosons) and the radiation–matter interaction

$$H_{\text{int}} = -\int \mathrm{d}\boldsymbol{r}\,\hat{\boldsymbol{P}}(\boldsymbol{r})\cdot\boldsymbol{E}(\boldsymbol{r}, t) \tag{3.76}$$

is a linear combination of the boson creation and annihilation operators via the polarization (induced dipole density) operator $\hat{\boldsymbol{P}}$. We neglect the A^2 term (2.31) in H_{int}, since we are interested in resonant processes.

Then the nth order nonlinear polarization is expressed as an n-fold time integral of an n-fold commutator of the polarization operator with H_{int} in the interaction representation, as shown in Sect. 2.6. What is peculiar in this idealized 'free boson' case is that both $\hat{\boldsymbol{P}}(t)$ and $H'(\tau)$ in the commutators are linear in the boson creation and annihilation operators. This means that the innermost commutator in (2.105) or (2.106) becomes a c-number, and thus, since the commutator of any operator with a c-number is zero, all the nonlinear susceptibilities in this idealized case vanish. In other words, a system of harmonic oscillators has no nonlinearity.

Realistic systems are generally more complex than independent atoms or free bosons. The origin of nonlinearity is precisely the deviation from the free boson character. This means either bosons with interaction or fermions with or without interaction. Thus, in electron systems, both Fermi statistics and the electron interaction cause nonlinearity. Excitons, each consisting of two fermions (an electron and a hole), are often approximated as bosons for low densities [56]. But there are interactions among them originating from the Coulomb interaction between the electrons. In view of the generally large coupling with the EM field, excitons are an interesting subject for the study of optical nonlinearity.

Excitons in an MS or NS system are a typical example of the excited states of a confined system extending coherently over the whole region of the sample, which is the very reason for a large optical nonlinearity. There have been proposals concerning the size enhancement of nonlinear (local) susceptibility, particularly the linear dependence of $\chi^{(3)}_{\mathrm{LWA}}$ on the size (length, area or volume) of the confined system under consideration [57, 58].

This proposal is a revival of the effect of 'giant oscillator strength' for a trapped exciton [59]. This concept was first proposed by Rashba and Gurgenishvili [60] and tested by Henry and Nassau [61] for excitons loosely bound by neutral impurities in semiconductors. The central idea is as follows. When the center-of-mass (CM) motion of an exciton is confined within a volume V_0, its transition dipole moment from the ground state is $\sqrt{V_0/\Omega}$ times that of a free exciton per unit cell $\bar{M}$. (Ω is the volume of a unit cell.)

The reasons for confinement can differ from case to case. In the first proposal, it was due to neutral impurities. The second application concerned the case of excitonic molecule formation from a single exciton state [62]. In this case, the confinement of the CM motion is due to the attraction of two excitons (with appropriate spin configurations). The third example of confinement of the CM motion is the sample size itself [57], i.e., the case of size quantization of the CM motion.

A simple example demonstrating this effect is a Frenkel exciton confined in a linear chain of N molecules. If we only allow the exciton to transfer between nearest neighbors, the size quantized wave numbers are

$$k = n\pi/(N+1)d\,, \quad (n = 1, 2, \ldots, N)\ ,$$

and the corresponding energies are $E_1(k_n) = E_0 - 2b\cos(k_n d)$, where E_0 is the site energy, $-b$ the transfer energy, and d the nearest neighbor distance. The n th normalized eigenstate is

$$|k_n\rangle = \left(\frac{2}{N+1}\right)^{1/2} \sum_j \sin(k_n j d)|j\,)\ , \tag{3.77}$$

where $|j\,)$ is the state in which only the j th molecule is excited. If we denote the dipole matrix element of a single molecule by M_0, that of the n th confined exciton state is

$$\langle k_n|\hat{P}|0\,) = M_0 \left(\frac{2}{N+1}\right)^{1/2} \sum_j \sin(k_n j d)\ , \tag{3.78}$$

where $|0\,)$ is the ground state. The last summation is equal to $\cot[n\pi/2(N+1)]$ for odd n and 0 for even n. When $n = 1$, this factor gives $\sim 2(N+1)/\pi$ if

$N + 1 \gg \pi$. Then altogether, $\langle k_1|\hat{P}|0\rangle \sim (N+1)^{1/2}M_0$, which shows the desired size enhancement effect.

Size enhancement of the transition dipole moment leads to the same for $\chi^{(3)}_{\mathrm{LWA}}$ along the following lines. (Note that we are concerned with the local susceptibility under LWA of systems with coherently extended excitations.) In LWA, we neglect the details of the position dependence of induced polarization. Therefore, $\chi^{(3)}_{\mathrm{LWA}}$ is defined as the induced dipole moment of the whole system divided by the volume of the sample. Among the transition dipole moments, there is one which is proportional to the square root of the sample volume (the size-enhanced one). The contribution of this matrix element to the linear susceptibility is volume independent, because the square of the matrix element has a linear volume dependence, and is cancelled out by the volume factor in the definition of induced polarization mentioned above. However, its contribution to the third-order susceptibility gives a linear volume dependence, because such a transition dipole moment can occur four times in the induced third-order polarization, giving a volume-squared dependence. Dividing it by the volume to define the induced polarization per unit volume, we get a volume-linear induced polarization [57,58]. This is the linear size (volume) dependence of $\chi^{(3)}_{\mathrm{LWA}}$. Since this dependence is connected with a particular level of the size quantized exciton, the effect should appear in the corresponding resonant processes.

The linear size enhancement of the particular transition dipole moment should not continue without limit. There must be a size region where the growth begins to saturate. There are two factors limiting enhancement. One is the validity limit of LWA, when the sample size becomes comparable to the wavelength λ of the resonant light, and the other is non-radiative scattering. The first factor is intrinsic, and the second extrinsic, depending on impurity concentration and/or temperature. The latter is often introduced in terms of the so-called coherent length (ξ_0) of the CM motion of an exciton, according to which saturation begins when the linear size of the sample L_{s} approaches $\min\{\xi_0, \lambda\}$ from below. It should be noted that the definition of the coherent length is not unique. It can be different for linear and nonlinear processes.

Let us first consider the intrinsic effect, i.e., the case $\xi_0 > \lambda$. This corresponds to the situation where non-radiative scatterings occur so seldom that the radiative width exceeds the non-radiative one. The amplitude of the coupling strength of an induced current density with the EM field is proportional to its spatial Fourier component with wave number q $(= \omega/c)$. [This is directly reflected in the radiative lifetime of this 'extended Lorentz oscillator', as can be seen from (3.39) in Sect. 3.5.] The matrix elements $\langle\mu|\hat{\boldsymbol{I}}(\boldsymbol{r})|\nu\rangle$ of the current density operator can have various spatial structures corresponding to the details of the matter wave functions. The structure consists of an atomic part and an envelope function. The latter is related to the size quantization of the CM motion of the exciton. When $L_{\mathrm{s}} \ll \lambda$, only the one with nodeless envelope function has the largest q-Fourier component, i.e., this state has the strongest interaction with the EM field. This is the LWA situation described by (3.78) in the case of a 1D Frenkel exciton.

The self-consistent EM field in such a small sized sample is almost uniform in space, and among the amplitudes $\{\tilde{F}_{\mu\nu}\}$ of the induced current density, defined in (2.75), only that for the spatially uniform vector potential makes a major contribution. In other words, the excited state with nodeless envelope function monopolizes the oscillator strength or the radiative width of the whole system.

As L_s becomes comparable with λ, the EM field of resonant light is obviously not spatially uniform in the sample. Then some of the q-Fourier component of the current density $\langle\mu|\boldsymbol{j}(\boldsymbol{r})|\nu\rangle$ can become appreciable even for non-uniform envelope functions. This means that some of the optically inactive components in LWA become optically active as L_s gets larger. In this situation, the radiative width is no longer monopolized by the state with nodeless envelope function, but is distributed among several other states with spatially varying current densities.

Generally speaking, the saturation of the transition dipole moment with increasing sample size proceeds by distributing the strength of the radiation–matter interaction from the size enhanced component to those states which are forbidden in LWA. The size dependence of the radiative lifetime or width of the confined excitons under LWA has been observed and analyzed for quantum dots [63] and for thin slabs [64, 65], which will be discussed in Sect. 4.1. (The experimental result due to [63] shows the size enhancement of the radiative decay rate. Indeed, the samples in this study seem to satisfy the condition $\xi_0 > \lambda \geq L_s$. However, it is frustrating that the width of the size quantized exciton level observed in the size-selective luminescence excitation spectrum is much larger than the radiative width and that the saturation of the size enhancement occurs at a size much smaller than the wavelength of the resonant light.) There should be a corresponding effect in the nonlinear response, but neither theoretical nor experimental study has been carried out for the condition $\xi_0 > \lambda$. The resonant enhancement of nonlinear signals beyond LWA, to be discussed in detail in Chap. 5, involves completely different physics.

The saturation of the size-enhanced $\chi^{(3)}_{\mathrm{LWA}}$ for the condition $\xi_0 < \lambda$ occurs in quite a different manner to that mentioned above. The size enhancement starts to saturate when $L_s \leq \xi_0$ $(< \lambda)$, and therefore we may still use LWA. This means that the saturation effect of the transition dipole moment does not yet set in. Instead, it is the cancellation effect that plays a central role in this case [66, 67]. There are two steps to explain the saturation along these lines. The first step is to see what the cancellation is, and the second is to see how it works in a given model for the nonlinear medium, where the effect of non-radiative scatterings will be described.

The cancellation problem in the nonlinear susceptibility, occurring in both cases mentioned above, arises from the way it is calculated mathematically.

Let us take for example the general expression (2.122) for $\chi^{(3)}$ given in Sect. 2.6. This contains the three-fold commutators of polarization operators at different times. Usually, we expand the commutators into a sum of eight terms, integrating each term over multiple time variables, and take the sum of the eight integrated terms. Cancellation occurs at this last stage, where the principal terms cancel each other, and the remaining terms work as the susceptibility of the system. Concerning the size dependence, the cancelling terms have linear size (volume) enhancement, so that it gives a significant effect in the large size region, whether or not these terms cancel out. If there were a method for calculating the three-fold commutators first, and then carrying out the time integrations, there would be no cancellation problem. But there is not yet such a general method for calculating nonlinear susceptibility, except in trivial model cases.

The operators in the eight-fold commutators do not generally commute, and their ordering in each term of the expanded form of the commutator must therefore be handled carefully. A three-fold commutator gives eight terms in its expanded form:

$$\Big[\big[[a,b],c\big],d\Big] = abcd + dbac + dcab + cbad - dcba - cabd - bacd - dabc \,. \tag{3.79}$$

In our case, a, b, c, d correspond to the polarization operators in the interaction representation $\hat{P}'_{\bar{\xi}}(\boldsymbol{r},t)$, $\hat{P}'_{\xi}(\boldsymbol{r}_1,\tau_1)$, $\hat{P}'_{\eta}(\boldsymbol{r}_2,\tau_2)$, $\hat{P}'_{\zeta}(\boldsymbol{r}_3,\tau_3)$, respectively, where $\bar{\xi}, \xi, \eta, \zeta$ stand for the Cartesian coordinates. Let us denote the ground state expectation value of the eight terms in the above expansion by $A_1, A_2, A_3, A_4, B_1, B_2, B_3, B_4$ in order of appearance. Note that A_i and B_i $(i = 1, 2, 3, 4)$ have the reverse ordering of the four operators and different signs.

The time integration of these terms as in (2.117) leads to various energy denominators. For example, the contribution from A_1 is

$$\sum_{\lambda}\sum_{\mu}\sum_{\nu} \frac{\langle 0|\hat{P}_{\bar{\xi}}(\boldsymbol{r})|\lambda\rangle\langle\lambda|\hat{P}_{\xi}(\boldsymbol{r}_1)|\mu\rangle\langle\mu|\hat{P}_{\eta}(\boldsymbol{r}_2)|\nu\rangle\langle\nu|\hat{P}_{\zeta}(\boldsymbol{r}_3)|0\rangle}{(E_{\nu 0} - \hbar\Omega_3)(E_{\mu 0} - \hbar\Omega_2)(E_{\lambda 0} - \hbar\Omega_1)} \,, \tag{3.80}$$

where E_μ, etc., are the eigenvalues of H_0, $E_{\mu\nu} = E_\mu - E_\nu$, and

$$\Omega_1 = \omega_1 + \omega_2 + \omega_3 + 3\mathrm{i}\delta \,, \quad \Omega_2 = \omega_2 + \omega_3 + 2\mathrm{i}\delta \,, \quad \Omega_3 = \omega_3 + \mathrm{i}\delta \,. \tag{3.81}$$

There are three types of intermediate states, namely, $|\lambda\rangle$, $|\mu\rangle$, and $|\nu\rangle$. The second can be the ground state, among other things. Such a term has a special meaning in the context of size enhancement, because the four matrix elements of $\hat{P}$ in (3.80) can all be size enhanced when both of the states $|\lambda\rangle$ and $|\nu\rangle$ are chosen as the exciton state with the nodeless envelope function discussed above.

Let us divide the eight $\{A_i, B_i\}$ into two groups $\{A_s(n_i), B_s(n_i) : s = 1,2,3,4,\ n_i = 0,2\}$ (16 terms altogether) according to whether the second intermediate state is the ground state ($n_i = 0$) or doubly excited state ($n_i = 2$).

Note that $E_{\mu 0} = 0$ for the case n_i=0. Each of these 16 terms involves a characteristic energy denominator, as well as the ordering of the four polarizations at different sites. If we expect the complete cancellation of interatomic terms in the case of free atoms, it should occur among the group of terms with the same energy denominator because, if there is cancellation, it must occur for any frequencies. From this guiding principle, we can classify the 16 terms into the following six groups:

(a) $A_1(2) + B_3(2)$,
(b) $A_3(2) + B_1(2)$,
(c) $A_2(2) + B_1(0) + B_4(0)$,
(d) $A_1(0) + A_4(0) + B_2(2)$,
(e) $A_4(2) + B_2(0) + B_3(0)$,
(f) $A_2(0) + A_3(0) + B_4(2)$.

In the case of free bosons, the cancellation is complete in each group. In the case of free atoms, although there are finite contributions from the doubly excited states in each of the $n_i = 2$ terms, the interatomic terms cancel out completely in each of the 6 groups, leaving only the independent contributions from single atoms, as shown below.

Let us consider an assembly of free atoms of a single species. It is assumed that the positions of the atoms are fixed, and each atom has only one excited state. All the polarizations of the light fields are assumed to be the same for simplicity, and we drop the indices $\bar{\xi}, \xi, \eta, \zeta$, etc. The doubly excited states ($n_i = 2$) mentioned above correspond to the states where two different atoms are excited. All the one- and two-atom excited states have common eigenenergies $E = E_1$ and $E = 2E_1$, respectively. The energy denominator of each group is given in this case by

(a) $(E_1 - \Omega_1)(2E_1 - \Omega_2)(E_1 - \Omega_3)$,
(b) $(E_1 + \Omega_1)(2E_1 + \Omega_2)(E_1 + \Omega_3)$,
(c) $(E_1 + \Omega_1)\ \Omega_2\ (E_1 + \Omega_3)$,
(d) $-(E_1 - \Omega_1)\ \Omega_2\ (E_1 - \Omega_3)$,
(e) $-(E_1 + \Omega_1)\ \Omega_2\ (E_1 - \Omega_3)$,
(f) $(E_1 - \Omega_1)\ \Omega_2\ (E_1 + \Omega_3)$.

In this model, finite polarization occurs only at atomic positions. We thus replace the position variables $(\boldsymbol{r}, \boldsymbol{r}_1, \boldsymbol{r}_2, \boldsymbol{r}_3)$ by the suffices (j, ℓ, m, n) to describe the atoms, with the understanding that position integrations are carried out around each atom. In order to see the consequence of Fermi statistics, we examine the restriction on the position variables in each term of a group. For example, $A_2(2)$ has matrix elements of the form

$$\langle 0|\hat{P}_n|\lambda\rangle\ \langle\lambda|\hat{P}_\ell|\mu\rangle\ \langle\mu|\hat{P}_j|\nu\rangle\ \langle\nu|\hat{P}_m|0\rangle\ , \tag{3.82}$$

where $|\mu\rangle$ is a two-atom excited state. In order for this term to be nonzero, we need to have either $n = j$, $\ell = m$ or $n = m$, $\ell = j$, but $j = \ell = m = n$ should be avoided. For $A_2(0)$, $|\mu\rangle$ is the ground state, so that we have $n = \ell$,

$j = m$ including the case $j = \ell = m = n$. In this manner, we can list the restrictions on the position variables for each term. Including the sign of each term, these restrictions exactly cancel the terms in each of the six groups of terms, except that the condition $j = \ell = m = n$ is allowed in the case $n_i = 0$, and forbidden in the case $n_i = 2$. The remaining contribution comes only from $j = \ell = m = n$ with $n_i = 0$, which means that the third order polarization at the j th atom occurs only from the same atom.

This is a local response, although we started from the general expression for $P^{(3)}$ as a nonlocal response. This is reasonable, because the wave functions in the model extend only within each atom. The value of $\chi^{(3)}_{\mathrm{LWA}}$ (per unit volume) is the sum of the terms with the denominators multiplied by a common factor $2M_0^4\rho_\mathrm{a}$, where M_0 is the dipole matrix element of an atom in this system and ρ_a the number density of the atoms. This is the $\chi^{(3)}$ of an atom times the density of the atoms, which is the expected result.

In the next step, we wish to ask what happens when the atomic excitations interact with one another. The simplest model to study this problem would be the 1D Frenkel excitons with a periodic boundary condition with period N [55, 66–68]. The size dependence is examined via the N dependence. The parameters in the model are the site energy E_0 and the transfer energy $-b$ of the Frenkel exciton, and the non-radiative damping γ. The interaction between excitons is neglected except for the Pauli principle, i.e., exclusion of the occupation of one site by more than two excitons. Because of the periodic boundary condition, the quantized values of k and the one-exciton eigenstates are slightly different from those given earlier in this section. We have

$$k = \frac{\ell\pi}{d} \quad (\ell = 1, 2, \ldots, N) , \tag{3.83}$$

$$|k\rangle = \frac{1}{\sqrt{N}} \sum_{\ell} \exp(\mathrm{i}k\ell d)|\ell\,\rangle , \tag{3.84}$$

respectively. The eigenstates of the two-exciton states are expressed as

$$|K, \kappa\rangle = \frac{2}{N} \sum_{\ell>m} \exp\left[\mathrm{i}K(\ell+m)d\right] \sin(\kappa|\ell-m|d)|\ell, m\,\rangle , \tag{3.85}$$

with energy eigenvalue

$$E_2(K, \kappa) = E_1(K+\kappa) + E_1(K-\kappa) , \tag{3.86}$$

where $E_1(k) = -2b\cos(kd)$ is the one-exciton energy. The periodic boundary condition required for the one- and two-exciton states leads to the allowed values of k, K, κ, viz.,

$$(k, K, \kappa) = (2n, 2\bar{m}, 2m' - 1)\frac{\pi d}{N} , \tag{3.87}$$

where $n, \bar{m} = 1, 2, \ldots, N$ and $m' = 1, 2, \ldots, (N-1)/2$. The quantization condition for (K, κ) depends on the even or odd nature of N, and the above condition is for odd N. Since a very similar analysis is possible for even N, we describe only the odd N case.

Using the analytic expressions for the dipole matrix elements at site ℓ,

$$\langle k|P_\ell|0\rangle = \frac{M_0}{\sqrt{N}} \exp(-\mathrm{i}k\ell d) , \tag{3.88}$$

$$\langle K,\kappa|P_\ell|k\rangle = M_0 \frac{2}{\sqrt{N^3}} \frac{\exp\big[-\mathrm{i}(2K-k)\ell d\big]\sin(\kappa d)}{\cos(k-K)d - \cos(\kappa d)} , \tag{3.89}$$

we can evaluate each of the 16 terms $\{A_s(n_i), B_s(n_i)\}$ defined earlier in this section.

Cancellation occurs within the same groups described before. For example, the group $B_4(2) + A_3(0) + A_2(0)$ gives the following contribution to $\chi^{(3)}$, after integration in the complex κ plane:

$$\frac{2M_0^4}{N^3} \sum_{k,k',K} \frac{S(k,k',K)\mathrm{e}^{\mathrm{i}knd}\mathrm{e}^{\mathrm{i}(2K-k)jd}\mathrm{e}^{\mathrm{i}(2K-k')\ell d}\mathrm{e}^{-\mathrm{i}k'md}}{(\Omega_2 - E_{k'k})[\Omega_3 + E_1(k)]} \bar{A}(K,k,\Omega_1) , \tag{3.90}$$

where $E_{kk'} = E_1(k) - E_1(k')$ and

$$S(k,k',K) = \frac{\sqrt{[E_0 + 2b\cos(kd) - \Omega_1]^2 - 16b^2\cos^2(Kd)}}{[\Omega_1 - E_1(2K-k)][\Omega_1 - E_1(2K-k') - E_{k'k}]} . \tag{3.91}$$

The factor $\bar{A}(K,k,\Omega_1)$ is given by

$$\bar{A}(K,k,\Omega_1) = -\mathrm{i}\,\mathrm{sgn}\,(\bar{b})\ \tan\frac{N(\bar{a}+\mathrm{i}\bar{b})}{2} , \tag{3.92}$$

where

$$\bar{a} + \mathrm{i}\bar{b} = \cos^{-1}\frac{E_0 + 2b\cos(kd) - \Omega_1}{4b\cos(Kd)} , \quad (\bar{a} > 0) . \tag{3.93}$$

This is the only factor which can explicitly depend on N. [The factor $1/N^3$ in the expression (3.90) disappears in the integral expression for this term over the three wave numbers.] For small N, the factor $\tan[N(\bar{a}+\mathrm{i}\bar{b})/2]$ is seen, via Taylor expansion, to exhibit N-linear dependence. For large N, however, this factor tends to unity, as long as the imaginary part $\bar{b}$ is finite, which is the case for finite $\delta = 0^+$, or finite non-radiative damping. Moreover, for all other combinations of the terms, we can explicitly derive similar results showing N-linear behavior for small N, and N-independence for large N.

In this way, introducing the transfer effect for excited states leads to saturation of the linear size enhancement of $\chi^{(3)}$ for small sizes. It should be noted that this argument has been made for a general nonlocal dependence on source and polarization positions $\{n,\ell,j,m\}$, and also for general conditions concerning frequencies $\{\omega_1,\omega_2,\omega_3\}$, i.e., for either resonant or non-resonant

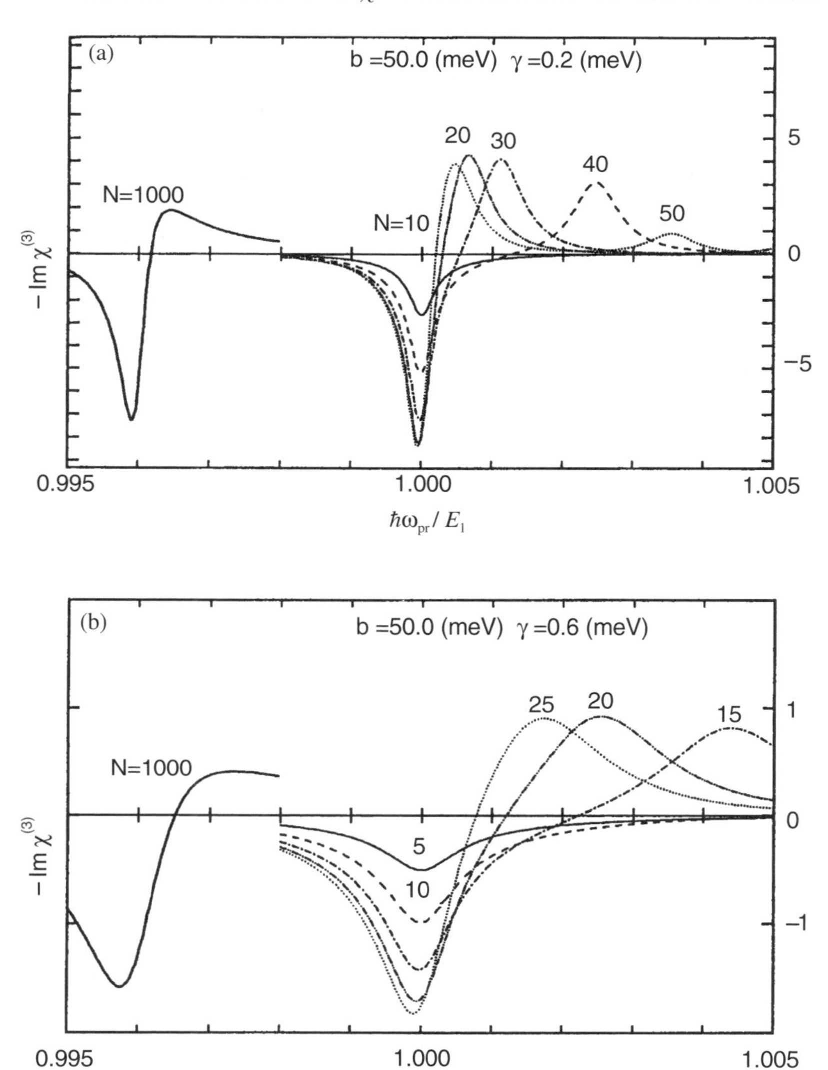

Fig. 3.1. Size (N) dependence of $\mathrm{Im}[\chi^{(3)}(\omega)]$ for the 1D Frenkel exciton in LWA for two γ values: (**a**) $\gamma = 0.2$ meV, (**b**) $\gamma = 0.6$ meV. *Numbers on curves* are N values. The curve for $N = 1\,000$ is shifted downward by 0.004 to avoid congestion. $E_1 = 1.0$ eV

conditions. Spano and Mukamel reported a similar demonstration of size dependence [69], claiming a better treatment than ours, in that they took radiative damping into account. However, it is wrong to put the effects of radiative damping into optical susceptibility, because it leads to double counting of the radiation–matter interaction, which is reflected in the response spectrum.

Finally, we discuss the effect of non-radiative damping in restricting the linear growth of $\chi^{(3)}$ within LWA [67]. For this purpose we can use the explicit evaluation of $\chi^{(3)}$ for the 1D Frenkel excitons described above. Finite non-radiative damping was introduced by simply keeping the parameter δ of the adiabatic switching finite, without letting it vanish. For the numerical calculation, we examined a pump–probe process, and considered the signal at the lowest exciton energy. The signal intensity was studied as a function of the transfer energy b, the non-radiative damping $\gamma(=\hbar\delta)$, and the size N.

Figure 3.1 shows $\mathrm{Im}[\chi^{(3)}]$ as a function of the probe frequency, with the pump frequency fixed at the lowest exciton energy, for various sizes N. The resonant structure near the lowest exciton consists of the combined shape of a lower energy dip and a higher energy peak, which represent the decrease in exciton absorption and the increase in the induced (one- to two-exciton) transition due to pumping. These transitions occur energetically very close to each other in the present model, but they are shifted by a small amount due to the different values of k, K, κ mentioned above. The magnitude of the dip is plotted as a function of N and b in Fig. 3.2.

This shows that the region of N-linear growth is dependent on b (and γ). Considering the similar curves for various γ values, we can derive the dependence of the saturation value of N as a function of b and γ, which turns out to be

$$N_{\mathrm{sat}} = \sqrt{\frac{6.72b}{\gamma}} \,. \tag{3.94}$$

This relation represents the ridge line in Fig. 3.2, i.e., below this value, $\mathrm{Im}[\chi^{(3)}]$ shows a linear size enhancement. In this way, we can derive the saturation behavior of the nonlinear susceptibility due to non-radiative damping in terms of a theory correctly treating the cancellation problem. The coherent length may be defined in this case as $N_{\mathrm{sat}}d$. Obviously, the level scheme of the two exciton states contributes to this value via the damping mechanism. If we want to define the 'coherent length' for the linear process, there will be a contribution from single exciton levels. This means that this coherent length differs from $N_{\mathrm{sat}}d$ mentioned above, as will be discussed in Sect. 4.1.2 in connection with Fig. 4.6. Although the dependence on $\sqrt{b/\gamma}$ turns out to be the same, their numerical coefficients are expected to be different.

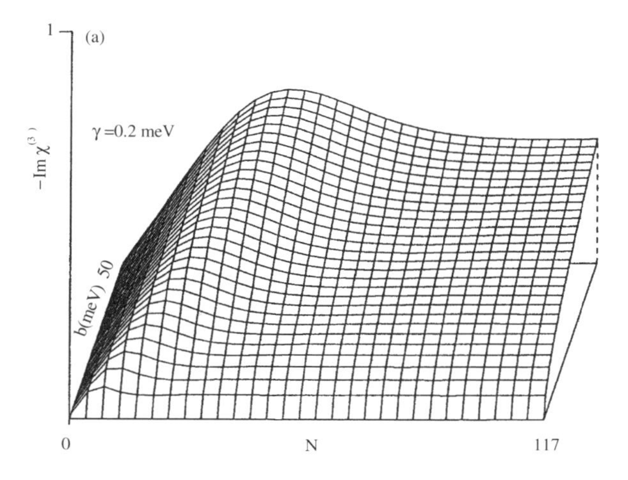

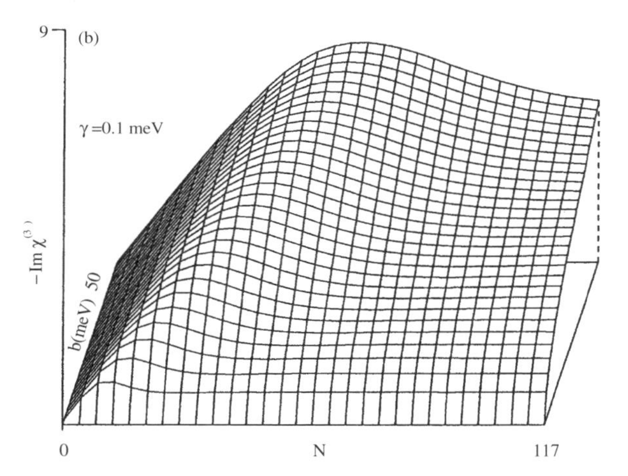

Fig. 3.2. Depth of the dip at $\hbar\omega_{\rm pr}/E_1 = 1.0$ in Fig. 3.1 as a function of N and b for two γ values: (**a**) $\gamma = 0.2$ meV, (**b**) $\gamma = 0.1$ meV

4. Application: Linear Response

In this chapter, the microscopic nonlocal response scheme is applied to various examples of linear response, which represent a characteristic feature of MS or NS systems, i.e., the coherence of matter wave functions leading to the nonlocality of response. This feature is mainly observed in MS or NS systems, but as some examples show, it can also be seen in certain bulk systems.

The fundamental equations to be handled are the coupled linear equations $\mathbf{S}\tilde{\boldsymbol{F}} = \tilde{\boldsymbol{F}}^{(0)}$ in Sect. 2.5, or $\mathbf{S}_{\mathrm{x}}\tilde{\boldsymbol{X}} = \tilde{\boldsymbol{F}}^{(0)}$ in Sect. 3.1, which determine the expansion coefficients of the current density and vector potential as in (2.80) and (2.83) for $\{\tilde{F}_{\mu\nu}\}$, or as in (3.6) and (3.7) for $\{\tilde{X}_{\mu\nu}\}$. This provides the necessary and sufficient information to fix the whole response. As discussed in Chap. 3, an essential role is played by the radiative correction included in these equations and by the size, shape and ω dependence of $\{\tilde{F}_{\mu\nu}\}$ or $\{\tilde{X}_{\mu\nu}\}$.

For the size-dependence argument, we generally need to consider three different length scales, i.e., the wavelength λ of light, sample size L_{s}, and coherence length ξ_0 due to non-radiative scattering. Various physical situations arise from differences in their relative magnitudes. Among them, we concentrate on the case $\xi_0 > \lambda, L_{\mathrm{s}}$, because it is the case where the coherence of matter extends over the whole sample, i.e., the case leading to the size, shape, and internal structure dependence of the optical response of mesoscopic systems.

4.1 Size Dependent Response

One of the main reasons for the interest in MS or NS systems is the possibility of achieving a dramatic increase in the coupling between matter and radiation as we change the size or shape of a sample and/or the light frequency. This increase can be seen in

(a) the oscillator strength (or coupling matrix element),
(b) the internal EM field.

These effects are based on different mechanisms. The former only contains information about matter wave functions, while the latter, arising from the self-consistently determined EM field, depends resonantly on sample size and light frequency.

The physics of (a) arises from the idea of giant oscillator strength, as mentioned in Sect. 3.9. The physical background of this effect is the (quasi) monopoly of oscillator strength by a single state among all the size-quantized excited states of the system. This state has a transition dipole density with least spatial variation, i.e., with no node in the envelope function, which favors the monopoly of oscillator strength in the long wavelength approximation (LWA). Since each atom (or unit structure) in a sample contributes to the transition dipole moment, the monopolized transition dipole density is of the order of Nf_1, where N is the number of constituent atoms and f_1 is the oscillator strength of a single atom. As sample size becomes larger, the validity of LWA diminishes and the linear size (N) dependence of the monopolized oscillator strength turns to be saturated. Since the total oscillator strength Nf_1 must be distributed among the size-quantized states, the excess part due to saturation of the monopolized component is taken up by some other states. The magnitude of the oscillator strength is reflected in the intensity and also in the radiative width of the response signal in typical far-field measurements. However, in a certain type of near-field measurement, a different situation can arise, as will be discussed in Sect. 4.3.2.

The mechanism (b) becomes important for resonant processes, because the resonance condition allows a strong contribution to the response field from a particular excited state. The condition for the large coupling between matter and radiation is exactly that for the occurrence of self-sustaining (SS) modes discussed in Sect. 3.2, i.e., $\det|\mathbf{S}| = 0$ or $\det|\mathbf{S}_x| = 0$. By changing the size and shape of a sample or the frequency of the incident light, we may expect a resonant enhancement of a particular component of $\{\tilde{F}_{\mu\nu}\}$ or $\{\tilde{X}_{\mu\nu}\}$, i.e., an internal field with particular spatial variation. In other words, this is a selective enhancement of the radiation–matter coupling. In linear response, this is an interference effect among various light waves in a nonlocal medium, a generalized version of Fabry–Pérot interference. In nonlinear processes, however, there is no known analogous effect, and it provides a new guiding principle for producing strong nonlinear signals [70,71]. The condition $\det|\mathbf{S}| = 0$ is a generalized phase-matching condition, when we seek suitable conditions for preparing an optimal radiation–matter coupling. What is different from the usual phase-matching condition is that we make full use of nonlocality in resonant processes, which involves the spatial structure of the induced current density and the self-consistently determined EM field, and contains the contribution of the resonant state with a special weight.

In the rest of this section, we will discuss the model systems:

- 1D, 2D and 3D lattices of N particles,
- a single slab,
- a single sphere,
- finite lattices with resonant Bragg scattering.

4.1.1 N 'Atoms' in 1D, 2D and 3D Arrangements

In order to discuss the size, shape, and internal structure dependence of the optical response of mesoscopic systems, we may think of various models. Among them, we consider simple systems of Frenkel excitons on finite lattices. By changing the total number N of 'atoms' and their rearrangements, we can systematically study a certain type of size, shape, and internal structure dependence. The atoms in these arrangements need not be real atoms: they can be molecules or quantum dots (QD) with a single excited level isolated from other levels, so that we may consider a system of Frenkel excitons on their finite lattices.

The case $N = 1$ corresponds to a single two-level system in vacuum, which was already treated in Sect. 3.5. For ω close to the resonance (i.e., in the rotating wave approximation), the coupled equations for $\{\tilde{X}_{\mu\nu}\}$ take the form

$$(E_{\mathrm{a}} - \hbar\omega + \mathcal{A}_{\mu 0,0\mu})\tilde{X}_{\mu 0} = \tilde{F}^{(0)}_{\mu 0} , \tag{4.1}$$

where $E_{\mathrm{a}} = E_{\mu} - E_0$, with suffices μ and 0 for the excited and ground states, respectively, and $\mathcal{A}_{\mu 0,0\mu}$ is the radiative correction (2.93) of Sect. 2.5. It was shown explicitly that $\mathrm{Im}[\mathcal{A}_{\mu 0,0\mu}]$ can be rewritten in exactly the same form as the one from the golden rule in QED, without even using LWA. [The expression in LWA, viz., (3.42) of Sect. 3.5, is a familiar expression.] In this model, $\mathrm{Re}[\mathcal{A}_{\mu 0,0\mu}]$ is a constant depending on the details of the relevant wave functions.

Now we consider arrays of N atoms, where atomic excitation can propagate from site to site as a Frenkel exciton via the dipole–dipole interaction H_{dd} among atoms. At the same time the size-quantized excitons interact via the transverse EM field, leading to the radiative shift and width of each exciton level. By diagonalizing H_{dd}, various components of the current densities associated with atomic excitations are transformed to those of size-quantized excitons. Since this transformation is unitary, the trace of any matrix is conserved by the transformation. This means that the trace of the radiative correction (matrix) is conserved before and after taking H_{dd} into account. In particular, the radiative width and shift as a whole are the same for the assemblies of non-interacting atoms and interacting atoms, which means that N times the radiative width (shift) of an atom is redistributed among the various size-quantized exciton states. What is most interesting about this redistribution is the monopolization by one (or several) of the exciton states under appropriate conditions, as discussed in Sect. 3.9 in connection with giant oscillator strength.

To be precise, we should note the following point in the above argument. The radiative correction $\tilde{A}$ is a function of frequency and the most general way of obtaining the radiative shifts and widths is to find the complex roots (ω) of $\det|\mathbf{S}| = 0$. A simpler approach is to take the diagonal part of $\mathcal{A}$ with its ω replaced by the diagonal (eigen)energy of the matter. This is acceptable when the level is isolated and the ω dependence of $\mathcal{A}$ is weak, which is usually the case. Taking this point into account, we should make a reservation concerning the above argument about conservation of the radiative correction. Indeed, it is true only when the redistribution of matter eigenenergies due to the introduction of a new interaction (H_{dd} in the above example) is not too wide in comparison with the original one (the atomic excitation energy). In the present case of atomic arrays, this is a perfectly acceptable condition, but the remark should be remembered in more general cases, e.g., the case treated in Sect. 4.1.4.

We consider finite lattices of 1D, 2D (square) and 3D (simple cubic) arrays of N atoms with lattice constant a_0 [72]. Each atom has an isolated resonant level with excitation energy E_{a}. The wavelength of resonant light λ $(= 2\pi\hbar c/E_{\mathrm{a}})$ is assumed to be much larger than a_0. For small N, the system size is much smaller than λ, so that LWA is valid. As N gets larger, the system size becomes comparable to, and eventually exceeds λ. Thus, a study of the N dependence of the response spectrum of this system will show us the evolution of the size enhancement of oscillator strength and its saturation.

For the numerical study, we used a spherical particle of semiconductor as 'atom', confining the center of mass (CM) motion of excitons. The material parameters are those for a CuCl [Z_3] exciton, with sphere radius $R = 15$ Å and and lattice constant $a_0 = 50$ Å. Since the exciton radius in bulk CuCl is about 7 Å, the picture of weak (CM) confinement is expected to remain valid in this sphere. Then we took the lowest level of the confined exciton as the resonant level of each atom ($E_{\mathrm{a}} = 3.2772$ eV), and the effect of all the remaining excited states was taken into account via a simplified background susceptibility treatment.

The resonant part of the transition dipole density is expressed in the form

$$\boldsymbol{\rho}_\xi^0(\boldsymbol{r}) = \hat{\boldsymbol{\rho}}_\xi^0 N_0 j_0(k_0 r)\Theta(R - r) \; , \tag{4.2}$$

$$N_0 = \sqrt{\pi/2R^3} \; , \tag{4.3}$$

where Θ is the Heaviside function, $j_0(k_0 r)$ is the 0th order spherical Bessel function, $\xi = x, y, z$ specifies the polarization direction, and $k_0 = \pi/R$. The amplitude of $\hat{\boldsymbol{\rho}}_\xi^0$ reflects the relative motion of the electron and hole in the exciton state and, in accordance with our picture of weak confinement, we ascribe to it the bulk value

$$|\hat{\boldsymbol{\rho}}_\xi^0|^2 = \frac{1}{4\pi}\Delta_{\mathrm{LT}}\epsilon_{\mathrm{b}} \; , \tag{4.4}$$

where ϵ_{b} and Δ_{LT} are the background dielectric constant and longitudinal–transverse splitting, respectively, in the bulk.

As for the non-resonant part of the induced polarization, we treat them as a whole in terms of a (constant) background susceptibility. We make a further simplifying assumption: we ascribe a single mode of polarization

$$\chi_{\xi\eta}^{\mathrm{b}}(\boldsymbol{r}, \boldsymbol{r}') = \chi_{\mathrm{b}}\delta_{\xi\eta}\Theta(R - r)\Theta(R - r') \; . \tag{4.5}$$

Generally speaking, the Heaviside function part should consist of various modes of polarization. However, as we wish to concentrate on the resonant spectrum of the system, this rather rough approximation will be permitted.

The matrix elements of H_{dd} are given in terms of the dipole moment of each sphere, i.e.,

$$\boldsymbol{\mu}_\xi = \int \mathrm{d}\boldsymbol{r}\boldsymbol{\rho}_\xi^0(\boldsymbol{r}) , \tag{4.6}$$

which are proportional to $R^{3/2}$. For a pair of atoms with transition dipole moments $\boldsymbol{\mu}_1(\parallel \xi), \boldsymbol{\mu}_2(\parallel \eta)$ separated by the vector $\boldsymbol{r}$, the radiative correction (2.93) is evaluated as

$$\begin{aligned}\mathcal{A}_{\xi 0,0\eta} = & -q^2\frac{\mathrm{e}^{\mathrm{i}qr}}{r}V_\mathrm{s}(qR)^2\Big\{\boldsymbol{\mu}_1\cdot\boldsymbol{\mu}_2^* - (\boldsymbol{\mu}_1\cdot\boldsymbol{r})(\boldsymbol{\mu}_2^*\cdot\boldsymbol{r}) \\ & -\frac{1-\mathrm{i}qr}{q^2r^2}\left[\boldsymbol{\mu}_1\cdot\boldsymbol{\mu}_2^* - 3(\boldsymbol{\mu}_1\cdot\boldsymbol{r})(\boldsymbol{\mu}_2^*\cdot\boldsymbol{r})\right]\Big\} \\ & -\frac{1}{r^3}\left[\boldsymbol{\mu}_1\cdot\boldsymbol{\mu}_2^* - 3(\boldsymbol{\mu}_1\cdot\boldsymbol{r})(\boldsymbol{\mu}_2^*\cdot\boldsymbol{r})\right] ,\end{aligned} \tag{4.7}$$

where the factor

$$V_\mathrm{s}(X) = \frac{\sin X}{X}\frac{\pi^2}{\pi^2 - X^2} \tag{4.8}$$

takes care of the finite spatial extent of the dipoles. When $r = 0$, i.e., for dipoles on the same sphere, we have the expression

$$\mathcal{A}_{\xi 0,0\eta} = -\boldsymbol{\mu}_1\cdot\boldsymbol{\mu}_2^*\frac{2q^2}{3R}\,\frac{\pi^2}{\pi^2-(qR)^2}\left[\mathrm{e}^{\mathrm{i}qR}V_\mathrm{s}(qR) + \frac{1}{2}\right] . \tag{4.9}$$

Figure 4.1a and b shows the far-field intensity resonantly scattered by 1D chains of various sizes for incident light polarized parallel to the chain (parallel to the z-axis). The angles of incident and scattered light are both perpendicular to the chain. The broad peak in each frame corresponds to the lowest level of quantized exciton states, whose envelope function has no node on the chain. Since the wavelength of the resonant light is about 4 000 Å in vacuum, LWA is valid in Fig. 4.1a but not in Fig. 4.1b. For small N values in Fig. 4.1a, no other peaks are seen across the whole spectral range. Hence, the lowest state monopolizes the oscillator strength, which is reflected in the fact that the (radiative) width of the broad peak increases linearly with N.

As N becomes larger (Fig. 4.1b), the width of the lowest energy peak gets saturated. At the same time a new peak starts to appear, and its width increases as N gets larger. The vertical dotted lines in Fig. 4.1b show the size-quantized energies of the exciton. If we number them 1, 2, 3, etc. in the direction of increasing energy, the peaks in the spectra correspond to odd quantum numbers. This is a consequence of the geometry, i.e., the polarization parallel to the z-axis and the incident direction perpendicular to the z-axis. Because the states with even quantum numbers have no transition

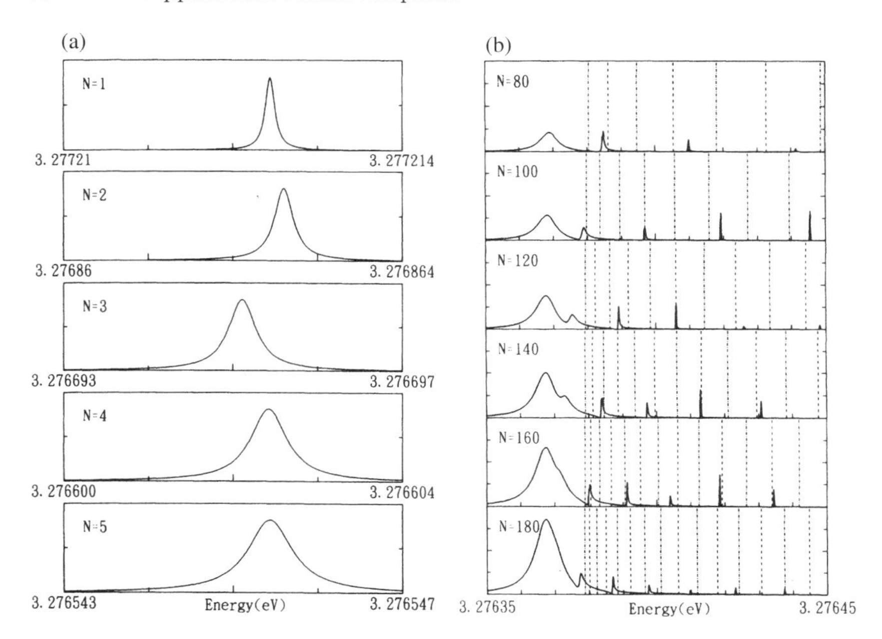

Fig. 4.1. Spectrum of scattered light for (**a**) small N values, and (**b**) large N values. *Vertical lines* in (**b**) indicate positions of size-quantized excited states [72]

dipole moment, they cannot be excited by incident light with uniform electric field in the z-direction.

The N dependence of the radiative shift and width of each exciton level is shown in Fig. 4.2a and b for polarization parallel and perpendicular to the z-axis. It is clear that enhancement and then saturation occur one after another for every state. It should be noted that states with even parity, which do not appear in the response spectrum, have similar enhancement of radiative width. In the LWA regime, they have much narrower widths than dipole active states. However, the dipole forbidden states which extend beyond λ, can couple strongly to multipole components of the EM field. This allows them to have large radiative widths, while their radiating fields have stronger spatial decay than the dipole active states. It is commonly believed that a large radiative decay rate for a state automatically implies an electric dipole character for radiation emitted by this state. But the present example shows that this is only true in LWA.

The size-quantized exciton levels in the geometries of Fig. 4.2a and b have a different ordering. This is understood as follows. Since H_{dd} is inversely proportional to the third power of interatomic distance, it can essentially be treated as a nearest-neighbor interaction in a 1D system. This leads to the energy eigenvalues

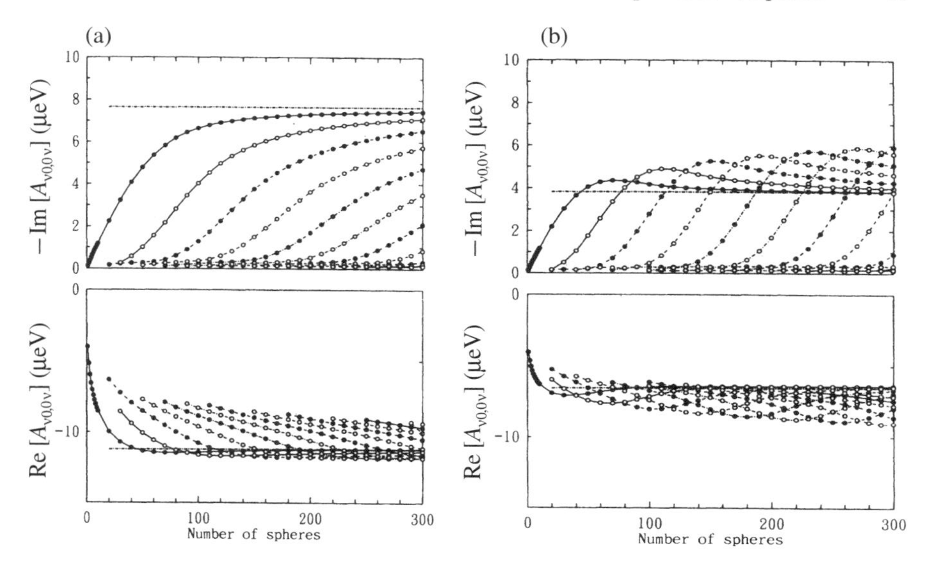

Fig. 4.2. N dependence of the radiative shift and width of each size-quantized state in a 1D lattice with polarization (**a**) parallel and (**b**) perpendicular to the chain. *Filled* and *open circles* correspond to modes with even and odd parities, respectively. The *horizontal line* in each frame indicates the value of the mode with $k = 0$ for $N = \infty$

$$E(k_n) = E_{\mathrm{a}} + \frac{2\mu_n^2(1 - 3\cos^2\theta)}{a_0^3}\cos(k_n a_0) , \tag{4.10}$$

where $k_n = n\pi/(N+1)a_0$ $(n = 1, 2, \ldots, N)$ and

$$\cos\theta = \frac{\boldsymbol{\mu}_n \cdot \boldsymbol{a}_0}{\mu_n a_0} , \quad \boldsymbol{\mu}_n = \int \mathrm{d}\boldsymbol{r} \langle k_n | \hat{\boldsymbol{P}}(\boldsymbol{r}) | g \rangle . \tag{4.11}$$

In the bulk ($N \to \infty$), there are three bands for x, y, and z polarization, for continuous wave number k. The size-quantized energies for a finite chain lie on these curves at every k_n. The curves are monotonically decreasing (increasing) functions of k for x and y (z) polarizations, and this causes the different orderings of the quantized levels for the two polarizations.

Although the results in Figs. 4.1 and 4.2 were all obtained from full consideration of H_{dd}, the simplified picture of nearest-neighbor interaction works well in the 1D chain.

Figure 4.3a and b shows the results of radiative shifts and widths for the 2D square lattices with N lattice points. The linear size $a_0\sqrt{N}$ of the lattice in the figures is still in the LWA regime. The results are different for polarization perpendicular (Fig. 4.3a) and parallel (Fig. 4.3b) to the lattice plane. Here again, the radiative corrections for dipole inactive states begin to have comparable values to those for dipole active states. Indeed, these states

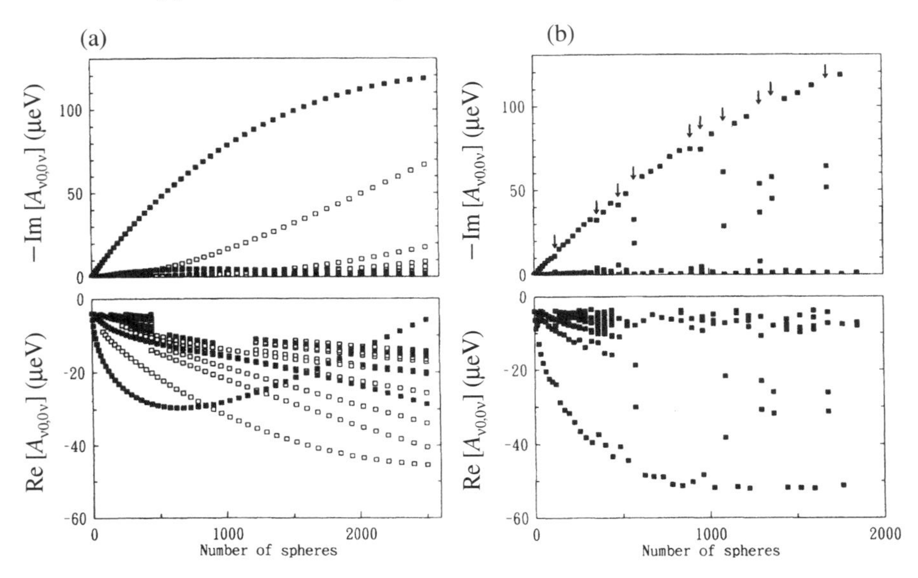

Fig. 4.3. N dependence of the radiative shift and width of each size-quantized state in a 2D square lattice with polarization (**a**) normal to the plane and (**b**) in the plane. *Filled* and *open squares* represent modes with and without the net dipole moment (in LWA), respectively

do not contribute to the far-field intensity, but have enough interaction with the multipole components of the near field.

There is some irregularity in Fig. 4.3b at sizes marked by downward arrows, where the monopoly of the radiative correction by a single state is broken. This is due to the appearance of nearly degenerate two-dipole active states at those values of N which share the radiative correction. This is considered to be a consequence of the long-range character of $H_{\rm dd}$, which is hidden in the 1D system, but becomes manifest in higher dimensions. If we truncate the farther neighbor interaction from $H_{\rm dd}$, these dip structures vanish.

The linear size of 3D simple cubic arrays cannot be very large in numerical calculations, so they all belong to the LWA regime. Once again, due to the long-range character of $H_{\rm dd}$ in the 3D lattice, the monopoly of oscillator strength ceases very quickly for $N > 3^3$, where many modes appear with comparable radiative widths.

Finally, we mention the limiting cases of the infinite 1D–3D lattices treated above, where analytic evaluation is possible. In each case, the quantum indices $(\mu\nu), (\tau\sigma)$ of the excitation or deexcitation contain 1D–3D wave vectors $(\bar{\boldsymbol{k}})$ as one of the quantum numbers. A component of the current density accompanying the transition $(\mu \leftarrow \nu)$ may be given as

$$\boldsymbol{I}_{\mu\nu,\bar{\boldsymbol{k}}}(\boldsymbol{r}) = \frac{1}{\sqrt{N_1}} \sum_{\boldsymbol{R}} \exp(\mathrm{i}\bar{\boldsymbol{k}}\cdot\boldsymbol{R}) \boldsymbol{I}^{(0)}_{\mu\nu}(\boldsymbol{r}-\boldsymbol{R}) \,, \tag{4.12}$$

where $\boldsymbol{I}^{(0)}_{\mu\nu}(\boldsymbol{r}-\boldsymbol{R})$ is the induced current density of an atom at site $\boldsymbol{R}$ and N_1 represents the number of the lattice points in the D-dimensional lattice. This definition of the wave vector means that, depending on whether we have excitation ($E_\mu > E_\nu$) or deexcitation ($E_\mu < E_\nu$), the wave vector $\bar{\boldsymbol{k}}$ is created or annihilated, respectively, in the matter. Then the radiative correction between the two current density components is written in terms of the Fourier transform as

$$\mathcal{A}_{\nu 0\bar{\boldsymbol{k}},0\nu\bar{\boldsymbol{k}}'} = \frac{-1}{2\pi^2 c^2} \int \mathrm{d}\boldsymbol{k} \frac{\tilde{\boldsymbol{I}}_{\nu 0,\bar{\boldsymbol{k}}}(-\boldsymbol{k})\cdot[\mathbf{1} - \hat{\boldsymbol{e}}_3(\boldsymbol{k})\hat{\boldsymbol{e}}_3(\boldsymbol{k})]\cdot\tilde{\boldsymbol{I}}_{0\nu,\bar{\boldsymbol{k}}'}(\boldsymbol{k})}{k^2 - (q+\mathrm{i}0^+)^2} \,, \tag{4.13}$$

where we substitute the appropriate expressions (4.12) for the current density for each of the 1D–3D cases. A more explicit form for the Fourier component is

$$\tilde{\boldsymbol{I}}_{\tau\sigma,\bar{\boldsymbol{k}}}(\boldsymbol{k}) = \int \mathrm{d}\boldsymbol{r}\, \boldsymbol{I}_{\tau\sigma,\bar{\boldsymbol{k}}}(\boldsymbol{r}) \mathrm{e}^{-\mathrm{i}\boldsymbol{k}\cdot\boldsymbol{r}} \,, \tag{4.14}$$

$$= \sqrt{N_1} \sum_{\boldsymbol{g}} \delta^{(D)}_{\boldsymbol{k},\bar{\boldsymbol{k}}+\boldsymbol{g}} \tilde{\boldsymbol{I}}^{(0)}_{\tau\sigma}(\boldsymbol{k}) \,, \tag{4.15}$$

where $D = 1,2,3$ is the dimensionality, $\boldsymbol{g}$ a reciprocal lattice vector of the D-dimensional lattice, and $\delta^{(D)}_{p,q}$ and $\delta^{(D)}(k)$ are the Kronecker delta and delta function in D dimensions, respectively.

Because of the translational symmetry in the 1D–3D arrays of atoms, $\mathcal{A}_{\nu 0\bar{\boldsymbol{k}},0\nu\bar{\boldsymbol{k}}'}$ is zero unless $\bar{\boldsymbol{k}} = -\bar{\boldsymbol{k}}'$, which is evident from the product $\delta^{(D)}_{-\boldsymbol{k},\bar{\boldsymbol{k}}+\boldsymbol{g}}\ \delta^{(D)}_{\boldsymbol{k},\bar{\boldsymbol{k}}'+\boldsymbol{g}'}$. Using

$$\delta^{(D)}_{\bar{\boldsymbol{k}}+\boldsymbol{k},\boldsymbol{g}}\ \delta^{(D)}_{\bar{\boldsymbol{k}}+\boldsymbol{k},\boldsymbol{g}'} = (\frac{2\pi}{La_0})^D\ \delta(\bar{\boldsymbol{k}}+\boldsymbol{k}-\boldsymbol{g})\ \delta^{(D)}_{\boldsymbol{g},\boldsymbol{g}'} \,, \quad (L^D = N_1) \tag{4.16}$$

and obtain the full form of the radiative correction

$$\mathcal{A}_{\nu 0\bar{\boldsymbol{k}},0\nu\bar{\boldsymbol{k}}'} = \delta_{\bar{\boldsymbol{k}},-\bar{\boldsymbol{k}}'} \frac{-1}{2\pi^2 c^2} \left(\frac{2\pi}{a_0}\right)^D \sum_{\boldsymbol{g}} \int \mathrm{d}\boldsymbol{k} \tag{4.17}$$

$$\frac{\delta^{(D)}(\bar{\boldsymbol{k}}+\boldsymbol{k}-\boldsymbol{g})}{k^2-(q+\mathrm{i}0^+)^2} \tilde{\boldsymbol{I}}^{(0)}_{0\nu}(-\boldsymbol{k})\cdot\left[\mathbf{1} - \hat{\boldsymbol{e}}_3(\boldsymbol{k})\hat{\boldsymbol{e}}_3(\boldsymbol{k})\right]\cdot\tilde{\boldsymbol{I}}^{(0)}_{\nu 0}(\boldsymbol{k}) \,.$$

Similarly, the matrix element of the dipole–dipole interaction is given by

$$\langle \nu 0\bar{\boldsymbol{k}}|H_{\mathrm{dd}}|0\nu\bar{\boldsymbol{k}}'\rangle = \delta_{\bar{\boldsymbol{k}},-\bar{\boldsymbol{k}}'} \frac{1}{2\pi^2} \left(\frac{2\pi}{a_0}\right)^D \sum_{\boldsymbol{g}} \int \mathrm{d}\boldsymbol{k} \frac{1}{k^2} \tag{4.18}$$

$$\delta^{(D)}(\bar{\boldsymbol{k}}+\boldsymbol{k}-\boldsymbol{g}) \tilde{\rho}^{(0)}_{0\nu}(-\boldsymbol{k}) \tilde{\rho}^{(0)}_{\nu 0}(\boldsymbol{k}) \,,$$

where $\tilde{\rho}^{(0)}_{\nu 0}(\boldsymbol{k})$ is the matrix element of the charge density operator, which is related to that of the current density operator through the continuity equation, i.e.,

$$\boldsymbol{k}\cdot\tilde{\boldsymbol{I}}^{(0)}_{\mu\nu}(\boldsymbol{k}) + \frac{1}{\hbar}(E_\mu - E_\nu)\tilde{\rho}^{(0)}_{\mu\nu}(\boldsymbol{k}) = 0 . \tag{4.19}$$

The excitation energy of the matter system is the sum of the on-site energy E_0 and H_{dd}, where the on-site part of the latter should be excluded. Thus the coefficient matrix of the linear equations $\mathbf{S}_{\mathrm{x}}\tilde{\boldsymbol{X}} = \tilde{\boldsymbol{F}}^{(0)}$ in Sect. 3.1 is given in the rotating wave approximation by

$$\mathbf{S}_{\mathrm{x}} = (E_0 - \hbar\omega)\mathbf{1} + H_{\mathrm{dd}} + \mathcal{A} . \tag{4.20}$$

The expression (4.17) is useful when discussing the dimensionality (D) dependence of the radiative width of a quantum state. It also applies to an isolated atom ($D = 0$), if we replace $\delta^{(D)}$ by 1 and omit the quantum number $\bar{k}$, which leads to (3.42) of Sect. 3.5 in LWA. This formula tells us that $\mathrm{Im}[\mathcal{A}]$ is nonzero only for the state with $\bar{\boldsymbol{k}}$ equal to the component of the wave number ω/c of the external light. Then $\mathrm{Im}[\mathcal{A}]$ becomes zero for $D = 3$, because $\bar{k} = \omega/c$ is not satisfied for general $\bar{\boldsymbol{k}}$. This corresponds to the situation where the light has no outer region to escape to in an infinite 3D matter system. For $D = 1$ and 2, only the states with $\bar{k}$ smaller than ω/c can have finite radiative widths, and the expression (4.17) indicates that they are enhanced relative to the $D = 0$ case by a factor $(2\pi q/a_0)^D = (\lambda/a_0)^D$, where λ is the wavelength of the resonant light. This means that the radiative decay rate is unevenly distributed among various $\bar{\boldsymbol{k}}$ states, i.e., it is monopolized by states with $\bar{k} \leq q$.

Figure 4.4a, b and c shows the energy dispersion of the dipole–dipole interaction in the 1D and 2D lattices [84]. It should be noted that the whole band width is larger by an order of magnitude in the 2D lattice than in the 1D lattice.

Figure 4.5a, b and c shows the radiative decay widths of the eigenstates of Fig. 4.4a, b and c [84]. Since the excitons with $\bar{k}$ larger than q cannot decay into light because of the translational symmetry, only those excitons with $\bar{k} < q$ have finite radiative widths. These values of the radiative width are calculated by replacing $\hbar\omega$ by the corresponding matter eigenvalue $E(\bar{k})$, rather than solving $\det|\mathbf{S}| = 0$ for its complex root. This approximation is permitted as long as $|\mathcal{A}|$ is small. The divergent part of $\mathrm{Im}\mathcal{A}$ near $\bar{k} \sim q$ in Fig. 4.5 is not therefore a good approximate solution of $\det|\mathbf{S}| = 0$. In this situation, $\det|\mathbf{S}| = 0$ should be solved. This leads to a converging result. This situation is discussed in detail by Orrit et al. [73] under the heading of 'a single pole approximation, and exact pole calculation' for the eigenvalue of the coupled system. Although their result is based on QED and ours on the semiclassical scheme, the results are exactly same.

4.1.2 Excitons in a Single Slab

This is a well studied case in the ABC problem for various thicknesses including a semi-infinite limit [51–53]. The resonant levels in such problems are

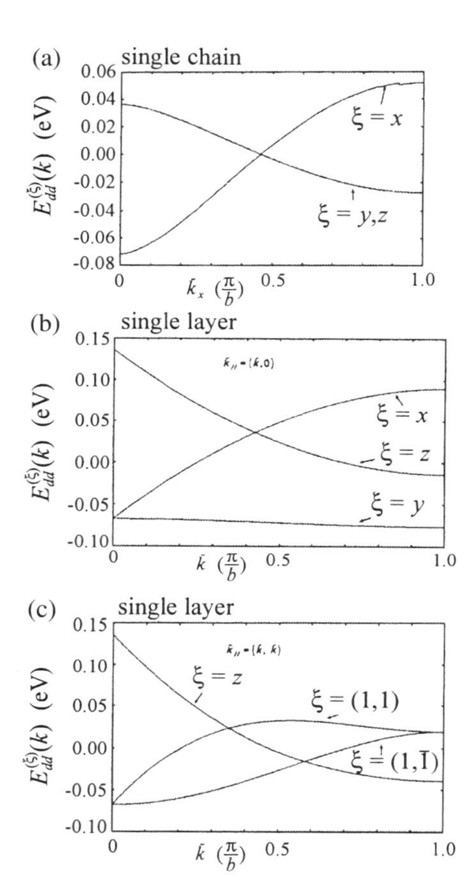

Fig. 4.4. Dispersion of Frenkel excitons with nearest neighbor dipole–dipole interaction for (**a**) 1D and (**b**, **c**) 2D lattices. For the 2D lattice, curves are shown for (**b**) [10] and (**c**) [11] directions of the wave vector

those of confined center-of-mass (CM) motion of excitons (weak confinement regime). When the thickness of the slab is roughly smaller than the diameter of exciton relative motion, confinement also affects the relative motion, leading to the so-called strong confinement regime, where the picture of individual confinement of the electron and hole is a better starting point than CM confinement in the weak confinement regime [39].

The standard form for the susceptibility in the ABC problem for a slab with normal incidence of light is

$$\chi(\boldsymbol{r}, \boldsymbol{r}'; \omega) = \chi_{\mathrm{b}} \delta(\boldsymbol{r} - \boldsymbol{r}') + \sum_{\nu} \bar{\chi}_{\nu}(\omega) I_{0\nu}(\boldsymbol{r}) I_{\nu 0}(\boldsymbol{r}') , \tag{4.21}$$

where $\{\nu\}$ refers to the size-quantized wave number K_ν perpendicular to the slab surface and

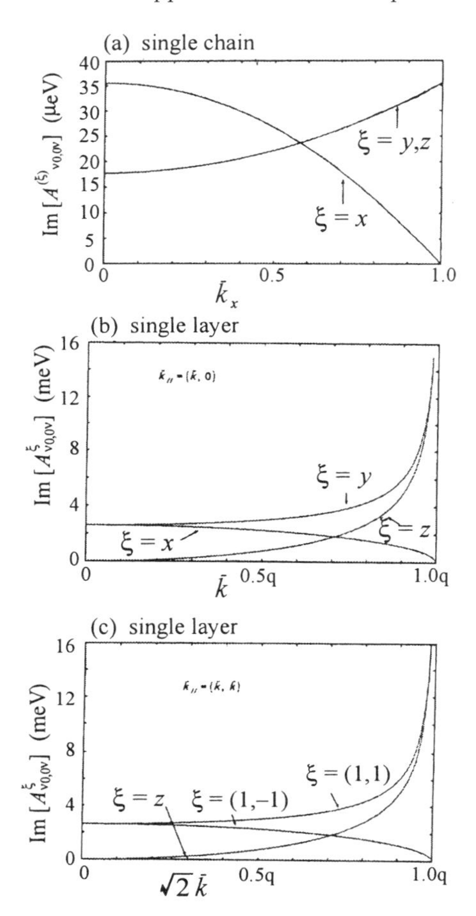

Fig. 4.5. Radiative widths of the exciton states in Fig. 4.4 as functions of the wave vector. (**a**) 1D chain, (**b**) 2D lattice, $\boldsymbol{k}_\parallel$ in the [10] direction, and (**c**) 2D lattice, $\boldsymbol{k}_\parallel$ in the [11] direction. $q = \omega/c$

$$\bar{\chi}_\nu(\omega) = \frac{B}{E_0 + (\hbar^2 K_\nu^2/2M_x) - \hbar\omega - \mathrm{i}\gamma_{\mathrm{nr}}} \tag{4.22}$$

represents the part in resonance with the size-quantized exciton levels. The background part χ_b takes care of the contribution from all the other matter states. The spatial variation of the induced current density (or polarization) depends on the model of the exciton wave function near the sample surface. One advanced microscopic models is the transition layer model of D'Andrea and Del Sole [51]. If we apply this model to a slab, it leads to the following form for ρ:

$$\rho_K(Z) = \frac{1}{\sqrt{2d}}\left[\exp^{-\mathrm{i}KZ} + R_K \exp^{\mathrm{i}KZ} + h_K(Z) + \bar{h}_K(\bar{Z})\right] , \tag{4.23}$$

where d is the thickness of the slab, $\bar{Z} = d - Z$, $R_K = (-P + \mathrm{i}K)/(P + \mathrm{i}K)$, $h_K = -(1 + R_K)\exp[-PZ]$, and $\bar{h}_K = -(1 + R_K)\exp[\mathrm{i}Kd - P\bar{Z}]$. The condition for size quantization depends on the microscopic model. The transition layer model for a slab gives [52]

$$K_\nu d - 2 \arctan \frac{K_\nu}{P} = \pi\nu\ , \quad (\nu = 1, 2, \dots)\ . \tag{4.24}$$

The transition layer model is a microscopic version of the dead layer model of Hopfield and Thomas [74], which takes into account the repulsion of excitons by the surface. The thickness of the layer is about $1/P$, and the value of P is evaluated from the model of the bulk exciton. Regarding P as a parameter, we can describe various situations. If $1/P = 0$, we have $K_\nu = \nu\pi/d$, which turns out to lead to the Pekar ABC. The transition layer model describes a more general situation than the Pekar ABC, giving a microscopic basis for the dead layer model.

As mentioned at the end of Sect. 3.8, the ABC problem for a slab can be solved in various forms which are actually equivalent, although they look rather different. The difference lies in the ways of treating resonant and non-resonant (background) polarization. The resonant part may be handled with or without ABC, and the non-resonant part either by the Maxwell boundary conditions (MBC) for the EM field or by the renormalized radiation Green function discussed in Sect. 3.4.2.

We do not give all the details of the different types of solution here, preferring to mention only their characteristic features. First of all, we discuss the radiative width of exciton levels. This is expected to grow with d as long as LWA is valid.

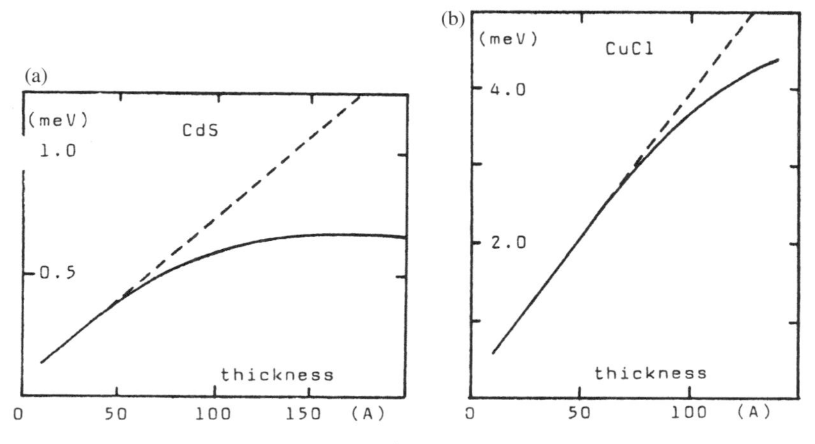

Fig. 4.6. Thickness dependence of the half width of the lowest exciton peak in the calculated absorption spectrum. Parameter values are for (**a**) CdS[A] and (**b**) CuCl[Z_3] excitons. Saturation of the linear growth is seen in both cases [65]

Figure 4.6 shows the d dependence of the width (FWHM) of the lowest exciton level in the absorption spectrum $A(\omega) = 1 - R(\omega) - T(\omega)$, where $R(\omega)$ and $T(\omega)$ are the reflectance and transmittance spectra, respectively. $A(\omega)$ is zero in the absence of non-radative damping γ_{nr}. For the 1s excitons in both CuCl [Z3] and CdS [A], the width exhibits linear size enhancement and a tendency to saturate as d approaches the coherent length $\xi = 2\pi/\mathrm{Im}[k_{\mathrm{c}}]$, where k_{c} is estimated from $E_0 + (\hbar^2 k_{\mathrm{c}}^2/2M_x) - \mathrm{i}\gamma_{\mathrm{nr}} - \hbar\omega = 0$. This behavior is consistent with that of the radiative width in general. It should also be noted that the width for $d \to 0$ is equal to the value of γ_{nr}, which means that the total width for general d is the sum of radiative and non-radiative widths. This guarantees that the susceptibility containing only non-radiative width produces the optical signal with radiative width via the nonlocal framework.

The above calculation of the spectral width was made using the ABC-free framework, which deals with the coupled equations of the boundary conditions $\mathbf{S}'\mathbf{E} = E_0\mathbf{a}$ discussed in Sect. 3.2. However, the same result should be obtained using the renormalized radiation Green function $\mathbf{G}_{\mathrm{r}}(\boldsymbol{r}, \boldsymbol{r}', \omega)$, following the revised version of the nonlocal response theory in Sect. 3.4.2. In particular, we calculate the interaction of the induced current density of the lowest ($\nu = 1$) size-quantized exciton with itself via the EM field, which contains the effect of the background dielectric constant ϵ_{b} of the slab. For normal incidence (parallel to the z-axis) of light polarized along the x-axis, we need only consider the x component of the current density, $j_1(Z)$, which depends only on Z. The renormalized Green function of the slab for $k_{\parallel} = 0$ (i.e., normal incidence) is the solution of the differential equation

$$\frac{\mathrm{d}^2 G_{\mathrm{r}}(Z, Z')}{\mathrm{d}Z^2} + q^2\left[1 + 4\pi\chi_{\mathrm{b}}\Theta(Z)\right]G_{\mathrm{r}}(Z, Z') = -4\pi\delta(Z - Z') \,, \tag{4.25}$$

where $\Theta(Z)$ is unity (zero) inside (outside) the slab. Its solution has the simple form

$$G_{\mathrm{r}}(Z, Z') = \frac{2\pi\mathrm{i}}{\bar{q}}\left\{\mathrm{e}^{\mathrm{i}\bar{q}|Z-Z'|} + \frac{2\delta^2\bar{g}^2}{1-\delta^2\bar{g}^2}\cos\left[\bar{q}(Z - Z')\right]\right. \\ \left. + \frac{\delta}{1-\delta^2\bar{g}^2}\left[\mathrm{e}^{\mathrm{i}\bar{q}(Z+Z')} + \mathrm{e}^{\mathrm{i}\bar{q}(2d-Z-Z')}\right]\right\} , \tag{4.26}$$

where $\bar{q} = q\sqrt{\epsilon_{\mathrm{b}}}$ with $\epsilon_{\mathrm{b}} = 1 + 4\pi\chi_{\mathrm{b}}$, $\delta = (\bar{q} - q)/(\bar{q} + q)$, $\bar{g} = \exp(\mathrm{i}\bar{q}d)$, and the source plane Z' is assumed to lie within the slab. The three terms on the right-hand side describe wave propagations directly from Z' to Z, via double reflection at surfaces $Z = 0$ and $Z = d$, and via single reflection at a surface, respectively. In the latter two propagations, repeated round trips are also included.

In terms of the radiation Green function, the radiative correction to the eigenenergy of the oscillator $j_1(Z)$ is given by

$$\mathcal{A}_{10,01}(\omega) = -\frac{1}{c^2}\int\int \mathrm{d}Z\,\mathrm{d}Z'\; j_1(Z)G_{\mathrm{r}}(Z, Z', \omega)\; j_1(Z') \,. \tag{4.27}$$

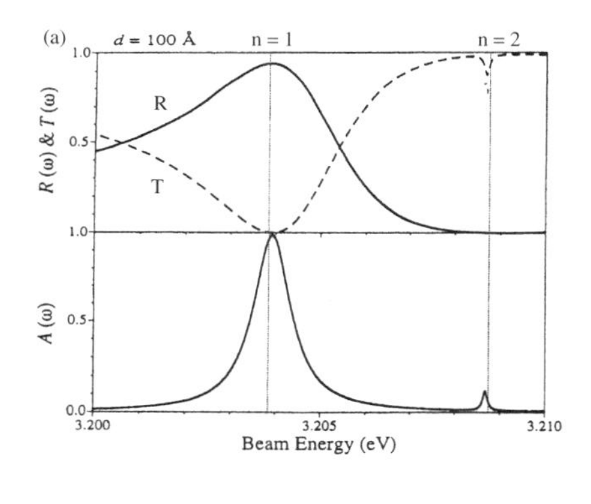

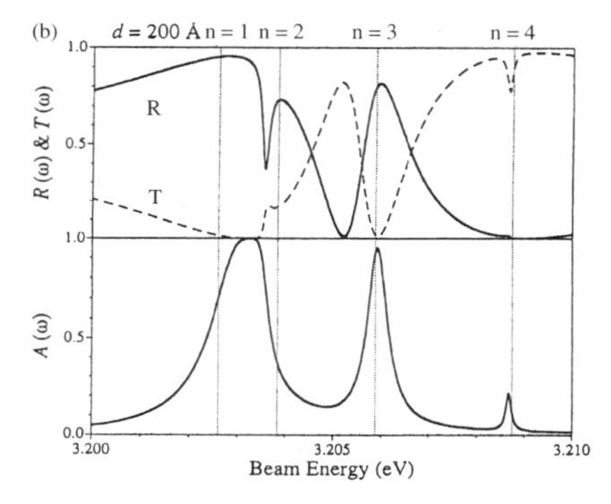

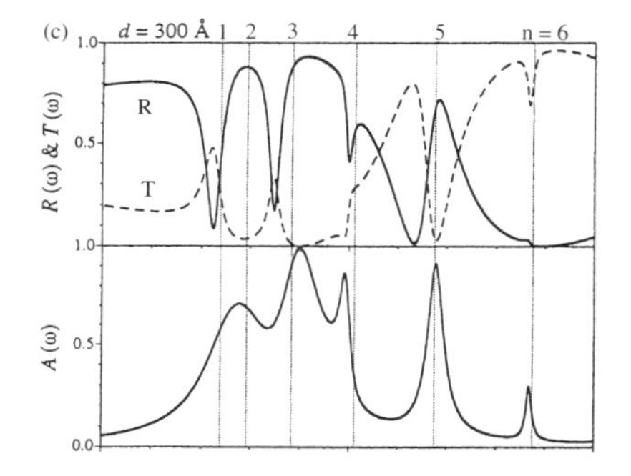

Fig. 4.7. Reflectance, transmittance and absorption spectra of a CuCl slab with thickness (**a**) 10, (**b**) 20, and (**c**) 30 nm. *Vertical dotted lines* show the position of the size-quantized exciton

Figure 4.7a, b and c shows an example of reflectance $R(\omega)$, transmittance $T(\omega)$, and absorption $A(\omega)$ for a slab of CuCl with various thicknesses, evaluated in terms of the radiative correction (4.27). The quantized levels of the exciton are also indicated by vertical dotted lines with quantum numbers n. For small thickness d, among the lower exciton levels, states with even quantum

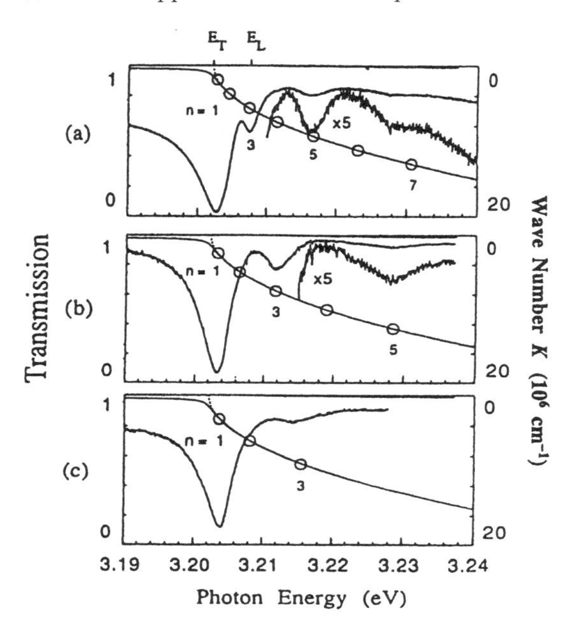

Fig. 4.8. Transmission spectra of CuCl thin films measured at 2 K for thickness (**a**) $d = 17.5 \pm 1.5$, (**b**) $d = 12.4 \pm 1.5$, and (**c**) $d = 9.7 \pm 1.5$ [64]. Polariton dispersion is shown by a *solid curve* with *open circles* for the energy positions of the size-quantized exciton. E_{T} and E_{L} are the energies of T and L bulk excitons at $K_{\mathrm{CM}} = 0$, respectively [64]

number ν ($k_\nu = \pi\nu/d$, $\nu = 1, 2, \ldots$) have much weaker structure than those with odd ν states. This is due to the dipole selection rule, which is valid in LWA. As d becomes comparable to the wavelength $2\pi c/(\omega\sqrt{\epsilon_{\mathrm{b}}})$ of the light in the slab, the contrast between the even and odd n states becomes weaker.

Figure 4.8 is an experimental demonstration of this dipole selection rule in the CuCl [Z_3] 1s exciton in thin slabs made by MBE (molecular beam epitaxy) growth [64, 75].

It is useful to see some details of the internal field produced by the incident light, especially for later use in the nonlinear response. For normally incident light with frequency ω on a slab of CuCl with thickness Na_0, where a_0 is the lattice constant, we determine the coefficients of the induced current density $\tilde{X}_{n0}(\omega)$ via $\tilde{\mathbf{X}} = (\mathbf{S}_{\mathrm{x}})^{-1}\tilde{\mathbf{F}}^{(0)}$, where n refers to the quantum number of the exciton CM motion [71].

Figure 4.9 shows the N and ω dependence of the factor $|\tilde{X}_{n0}(\omega)|^2$ for several n values, exhibiting its sensitive dependence on ω and the slab thickness Na_0. The ridges in the figure showing resonant enhancements correspond to the poles of $\det(\mathbf{S})_{\mathrm{x}}^{-1}$, and they cross the curves representing the size-quantized exciton energies $E_{\mathrm{x}n}$ as functions of N. Since $(\omega - E_{\mathrm{x}n}/\hbar)$ is one of the resonant factors in the triply resonant terms of $\chi^{(3)}$ for the pump–probe

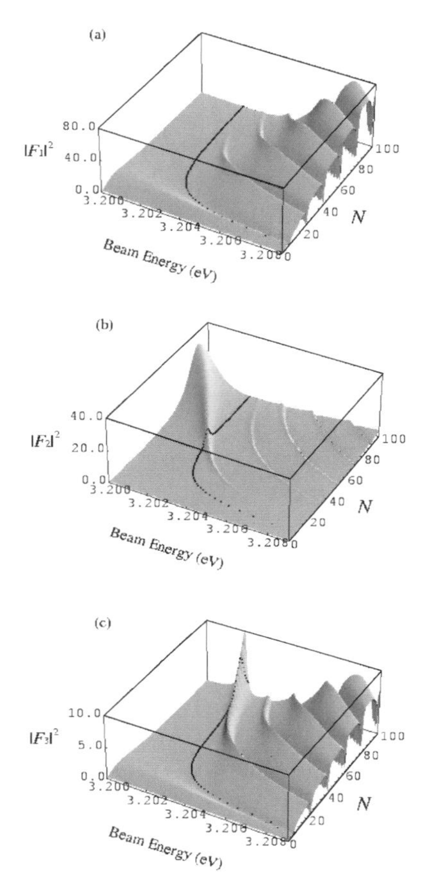

Fig. 4.9. Intensity of the induced electric field with spatial structure corresponding to the nth quantized state ($n = 1, 2, 3$) as a function of system size N and beam energy. The *dotted line* shows the cross-section of the surface at the energy of the $n = 1$ exciton for each N. $\gamma = 0.06$ meV [71]

process to be mentioned in Chap. 5, the crossing points are candidates for double enhancement with respect to frequency and size.

Figure 4.10 shows the absolute square of the internal field as a function of the position and frequency for various thicknesses. It should be noted that the spatial structure changes considerably with the thickness of the slab and the frequency. The characteristic structures are the reflection of the largest $\tilde{F}_{n0}$ for given N and ω. The wavelength of such a structure has nothing to do with that in vacuum ($2\pi c/\omega$), but is determined self-consistently with the resonantly excited matter state, which has wavelength much shorter than $2\pi c/\omega$. This feature is another view of the well-known Fabry–Pérot interfer-

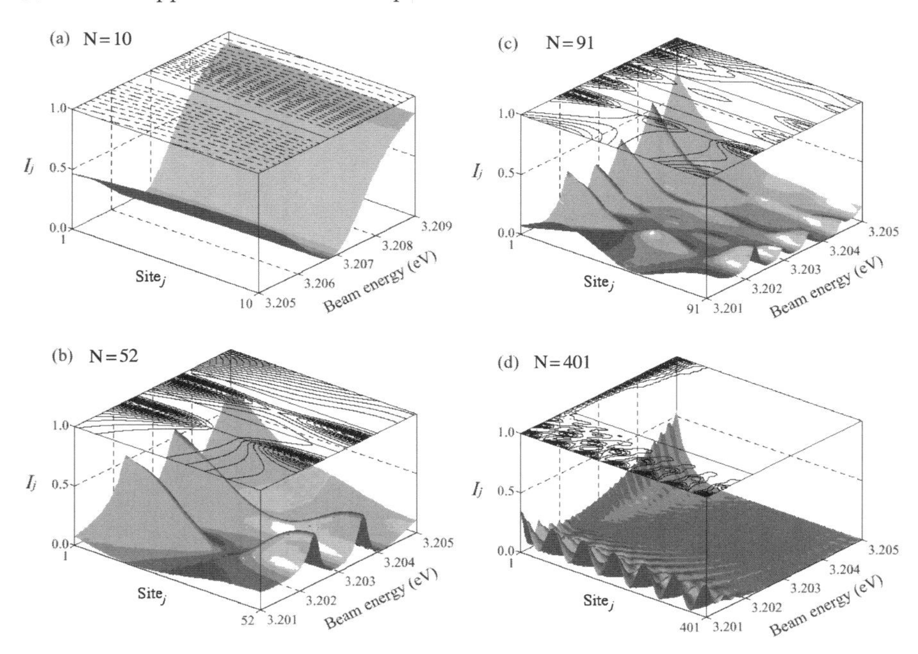

Fig. 4.10a–d. Normalized internal field pattern as a function of beam energy for $N = 10$, 52, 91 and 401. Parameter values are the same as those in Fig. 4.9 [71]

ence in the linear response, but it can lead to remarkable new effects in the nonlinear regime, as will be discussed in detail in Chap. 5.

Figure 4.11 shows the absorption spectrum of a CuCl slab (for d = 28.6 nm) and the thickness dependence of peak positions and widths caused by the radiative correction. The d dependence arises not only from the confined exciton states, but also from light confinement due to the background dielectric constant, whose effect is taken into account via the Green function $G_r(Z, Z')$ of (4.26).

4.1.3 Excitons in a Single Sphere

The optical response of a dielectric sphere has been studied extensively [76]. When the medium is described by a local dielectric constant ϵ_s, it is easy to solve the Maxwell equations both inside and outside the sphere, and the response is obtained by connecting the solutions via standard Maxwell boundary conditions (MBC). For a medium with frequency-dependent ϵ_s, the same procedure can be used to determine the optical response. This case typically includes the matter resonances due to impurities, defects or excitons with infinitely large mass, whose spatial distribution is regarded as macroscopically uniform.

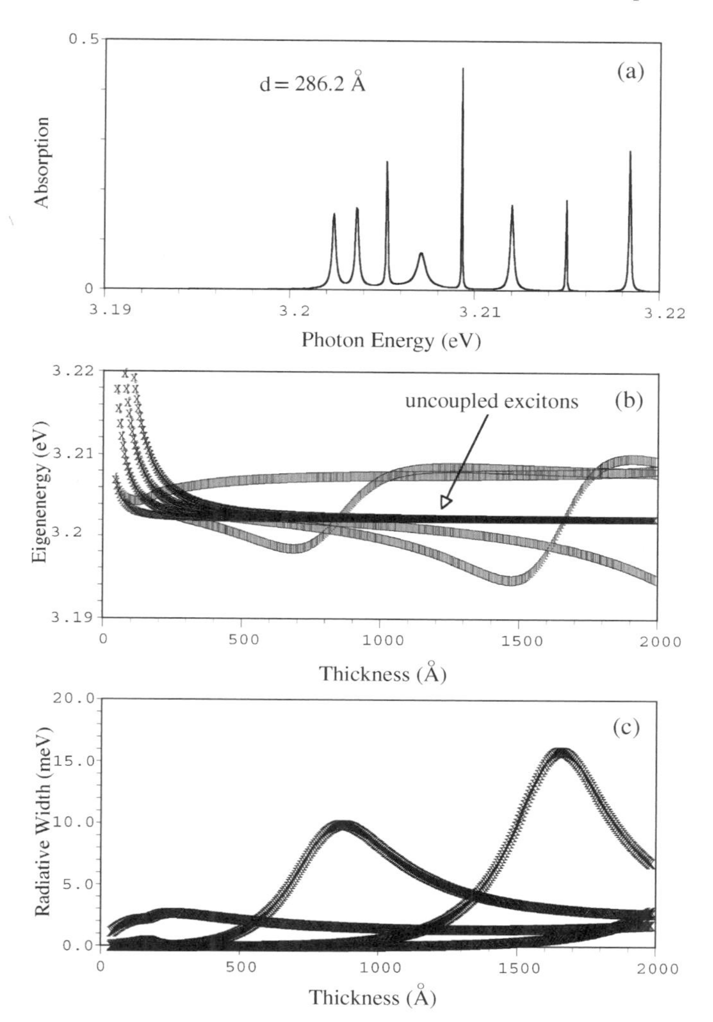

Fig. 4.11. (**a**) Absorption spectrum of a CuCl slab for $d = 28.6$ nm. Thickness dependence of (**b**) the peak positions and (**c**) the widths caused by the radiative correction

However, there are cases where we need to consider the spatial dependence of ϵ_s explicitly. A typical case is the confined exciton with finite mass. Such a medium is described by a nonlocal dielectric function $\epsilon(\boldsymbol{r}, \boldsymbol{r}'; \omega)$ or, in the bulk limit, by a spatially dispersive one $\epsilon(\boldsymbol{k}, \omega)$. The latter is the starting point for the ABC problem, as mentioned in Sect. 3.8.

The optical response of an exciton in a sphere was studied by Ekimov et al. [39] and by Ruppin [38] according to the standard procedure of macroscopic optical response using the MBC and the assumed ABC (of Pekar type). They were able to derive the response spectrum with resonance structure corresponding to size-quantized exciton levels. Such a spectrum could be directly compared with measurement, if a single-particle spectrum were available. Otherwise, a size-averaged spectrum is calculated to compare with the measurement for an assembly of fine particles [38]. In these works no information was derived about the L, T and LT-mixed character of each resonance and the amount of radiative shift and width of each resonance. In order to make these points clearer, it is more suitable to use the nonlocal response framework.

In the nonlocal response scenario, the optical response of an exciton in a sphere is calculated as follows. First, we solve the level structure of the exciton in the presence of the electron–hole (e–h) exchange interaction, which gives an additional energy to each quantized level according to its L, T, and LT-mixed character. In the second step, we consider the coupling of these exciton levels with the transverse EM field in the manner mentioned in Sect. 2.5. This gives us the response spectrum with a detailed knowledge of the radiative shift and width for each level and of the e–h exchange energy.

The size quantization of an exciton has weak and strong confinement regimes, as in the case of a slab mentioned in Sect. 4.1.2. Here we restrict ourselves to the case of weak confinement, where only the center-of-mass (CM) motion of the exciton is size-quantized with the relative motion being kept as in the bulk. (The case of strong confinement can be treated similarly via different expressions for the size quantization energy and induced dipole density.)

The induced dipole density accompanying the confined exciton can be described by (4.2), as in Sect. 4.1.1. The only difference is that we consider the multi-level structure of the exciton explicitly in this section, while only the lowest level was considered in Sect. 4.1.1. This is because here we are interested in the size variation of the level structure and its consequences for the response spectrum.

Because of the spherical symmetry, every state is specified by the quantum number J of the total angular momentum and the quantum number M of its z component. $\boldsymbol{J}$ is the sum of the orbital angular momentum (quantum number ℓ) of the confined CM motion of the exciton and the angular momentum of the dipole moment for the e–h relative motion [quantum number ℓ' $(= 1)$]. The induced polarization (dipole density) corresponding to a quantized exciton state has the form [37, 77]

$$\boldsymbol{P}_{nJ\ell M}(\boldsymbol{r}) = \mu\sqrt{\frac{2}{R^3}}\frac{j_\ell(k_{n\ell}r)}{j_{\ell+1}(k_{n\ell}R)}\mathbf{Y}_{J\ell M}(\hat{\boldsymbol{r}})\Theta(R-r)\ , \tag{4.28}$$

where R is the radius of the sphere, $j_\ell(x)$ the spherical Bessel function, and $k_{n\ell}$ the n th root of $j_\ell(k_{n\ell}R) = 0$. The vector spherical harmonics are defined by

$$\mathbf{Y}_{J\ell M}(\hat{\boldsymbol{r}}) = \sum_{m=-\ell}^{\ell}\sum_{s=-1}^{1}\langle \ell m 1 s|\ell 1 J M\rangle Y_{\ell m}(\hat{\boldsymbol{r}})\boldsymbol{e}_s\ , \tag{4.29}$$

in terms of the Clebsch–Gordan coefficient and the spherical harmonics $Y_{\ell m}(\hat{\boldsymbol{r}})$. The unit vectors $\boldsymbol{e}_s$ are defined in terms of those for Cartesian coordinates by

$$\boldsymbol{e}_{\pm 1} = \mp(\boldsymbol{x} + \mathrm{i}\boldsymbol{y})/\sqrt{2}\,, \quad \boldsymbol{e}_0 = \boldsymbol{z}\,. \tag{4.30}$$

For a fixed (1s-type) exciton relative motion, the quantum numbers n, ℓ and m of the confined state specify the radial and angular components of the CM motion. The quantity μ is related to the LT splitting of this exciton in the bulk by $|\mu|^2 = \epsilon_{\mathrm{b}}\Delta_{\mathrm{LT}}/4\pi$.

The interaction between components of the induced dipole density of the confined exciton is screened in two ways by the background dielectric constant of the sphere. One is the bulk-like screening, and the other is the surface (or image) charge effect. The former is written as the Coulomb interaction among the induced charge densities, which is screened by the background dielectric constant ϵ_b, viz.,

$$\langle\xi|H_{\mathrm{dd}}|\xi'\rangle = \int\!\!\int \mathrm{d}\boldsymbol{r}\mathrm{d}\boldsymbol{r}'\nabla\cdot\boldsymbol{P}_\xi(\boldsymbol{r})^* \frac{1}{\epsilon_{\mathrm{b}}|\boldsymbol{r}-\boldsymbol{r}'|}\nabla'\cdot\boldsymbol{P}_{\xi'}(\boldsymbol{r}')\,, \tag{4.31}$$

where ξ represents the set of quantum numbers $\{n, J, \ell, M\}$. The matrix element is obviously proportional to $\delta_{JJ'}\delta_{MM'}$ because of the spherical symmetry. The explicit expression for this matrix element is

$$\langle\xi|H_{\mathrm{dd}}|\xi'\rangle = \delta_{JJ'}\delta_{MM'}\Delta_{\mathrm{LT}}\begin{cases} P_{J\ell}^2\delta_{nn'} & \ell = \ell'\,, \\ \sqrt{J(J+1)}f_{nn'J} & \ell = J+1,\ \ell' = J-1\,, \\ \sqrt{J(J+1)}f_{n'nJ} & \ell = J-1,\ \ell' = J+1\,, \end{cases} \tag{4.32}$$

where

$$P_{J\ell}^2 = \begin{cases} J/(2J+1) & \ell = J-1\,, \\ 0 & \ell = J\,, \\ (J+1)/(2J+1) & \ell = J+1\,, \end{cases} \tag{4.33}$$

and

$$f_{nmJ} = \frac{2\kappa_{n,J+1}}{\kappa_{m,J-1}(\kappa_{n,J+1}^2 - \kappa_{m,J-1}^2)}\,, \quad (\kappa_{m,J} = k_{m,J}R)\,. \tag{4.34}$$

This interaction couples the different, as well as the same, angular momentum (ℓ, ℓ') states of the CM motion. From the above result, excitons with $J = \ell$ are pure T modes and do not contribute to the dipole–dipole interaction. This T-mode character can be shown more directly by calculating $\nabla\cdot\boldsymbol{P}_\xi$, which is proportional to the factor $P_{J\ell}$.

The other term of the dipole–dipole interaction affected by the surface charge density due to polarization of the background dielectric is evaluated as [37]

$$\langle\xi|H_{\mathrm{dd}}^{(\mathrm{SC})}|\xi'\rangle = \frac{(\epsilon_1-\epsilon_2)(J+1)}{\epsilon_2(J+1)+J}\frac{2}{\kappa_{nJ-1}\kappa_{n'J-1}}\Delta_{\mathrm{LT}}\delta_{JJ'}\delta_{MM'}\delta_{\ell,J-1}\delta_{\ell',J-1}\,, \tag{4.35}$$

where ϵ_1 and ϵ_2 are the background dielectric constants in the sphere and the surroundings, respectively. The size-quantized exciton levels in the presence of the e–h exchange interaction are obtained by diagonalizing the sum of the kinetic energy of CM motion, H_{dd}, and $H_{\mathrm{dd}}^{(\mathrm{SC})}$. Because of the spherical symmetry, this matrix is block diagonal with respect to J and M. Each block consists of three sub-blocks corresponding to $\ell = J-1, J, J+1$, and each sub-block is defined by the basis set of the radial CM motions with quantum number n.

Figure 4.12 shows an example of the size (R) dependence of the quantized exciton levels for $J = 1$, $\ell = (0, 2)$ [37]. The number of the basis functions for the radial component is 100 in this example. It should be noted that, in the bulk limit, all levels except one converge to the energies of the L and T modes of the bulk exciton at $K = 0$. The exception is the surface mode, whose energy is J dependent. Around the energy of the surface mode, every curve shows anti-crossing behavior, and an anomalous behavior with respect to the L or T character (see below), which should have certain consequences for the optical response spectrum.

The L, T, or LT-mixed character of the size-quantized levels is determined by calculating $\nabla\cdot\boldsymbol{P}$ and $\nabla\times\boldsymbol{P}$. If the former (latter) is zero everywhere, it is a T (L) mode, and otherwise, it is an LT-mixed mode. Since the T mode has no induced charge, it makes no contribution to the dipole–dipole interaction. All the states with solid curves in Fig. 4.12 are LT-mixed modes, although the degree of L and T character changes from level to level, and with the radius

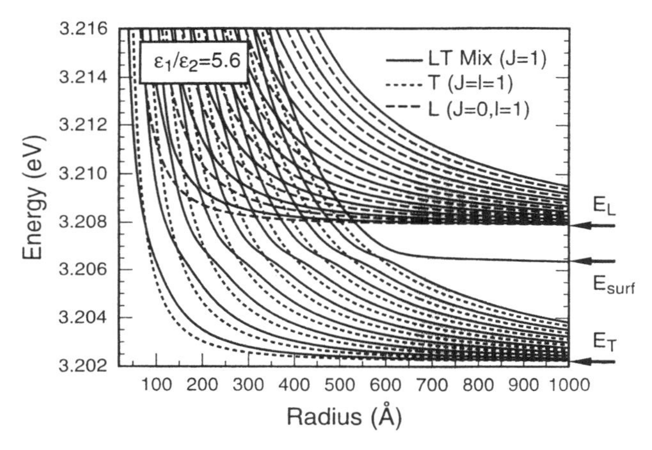

Fig. 4.12. Size dependence of the confined exciton energies with $J = 0$ and $J = 1$. Parameter values are those of the CuCl[Z_3] exciton. E_{surf} is the energy of the surface mode for the $J = 1$ exciton with infinitely heavy CM

R. On the other hand, the levels for $J = \ell \geq 1$ are all T modes, and their energies therefore consist only of the size-quantized CM kinetic energies.

Figure 4.13 shows the R dependence of the lowest exciton levels in the large R region. The lowest state is an LT-mixed mode and there is no parallel level in its neighborhood in contrast to the bulk T and L levels separated by Δ_{LT}.

The radiative correction $\mathcal{A}_{\xi 0, 0\xi'}$ is written in terms of the dipole density as

$$\mathcal{A}_{\xi 0,0\xi'}(\omega) = -q^2 \int\int \mathrm{d}\boldsymbol{r}\mathrm{d}\boldsymbol{r}' \boldsymbol{P}_\xi(\boldsymbol{r})^* \cdot \mathbf{G}_{\mathrm{r}}^{(\mathrm{M})}(\boldsymbol{r},\boldsymbol{r}';\omega) \cdot \boldsymbol{P}_{\xi'}(\boldsymbol{r}') \ , \tag{4.36}$$

where $\mathbf{G}_{\mathrm{r}}^{(\mathrm{M})}(\boldsymbol{r},\boldsymbol{r}';\omega)$ is the radiation Green function (tensor) defined in Sect. 3.4. Here again, the non-vanishing matrix element occurs only for $\ell - \ell' = 2, 0, -2$. Their explicit expressions are given in [78], and also included in [79]. For small R, $\mathrm{Im}[\mathcal{A}]$ is proportional to the oscillator strength of the confined exciton $|\xi\rangle$.

Figure 4.14 shows the R dependence of the exciton oscillator strength. Except for the small R region, the oscillator strength is concentrated at levels close to the surface mode. The lowest state changes from an LT-mixed level to a T level at $R = 7.6$ nm, which can also be seen from the oscillator strengths.

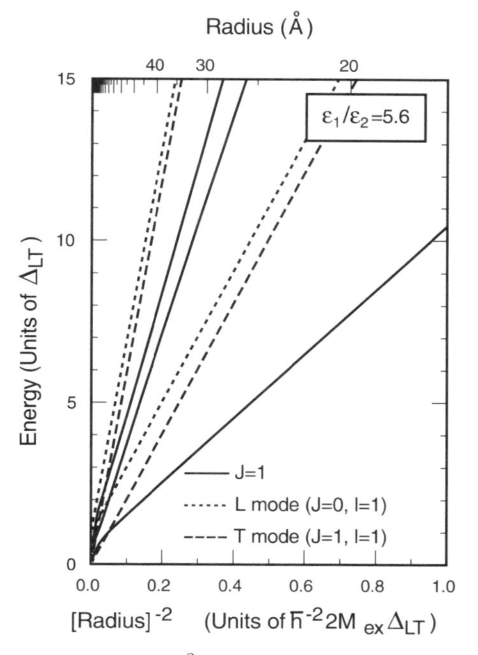

Fig. 4.13. $1/R^2$ plot of the confined energies of a few lowest energy members of Fig. 4.11 [37]

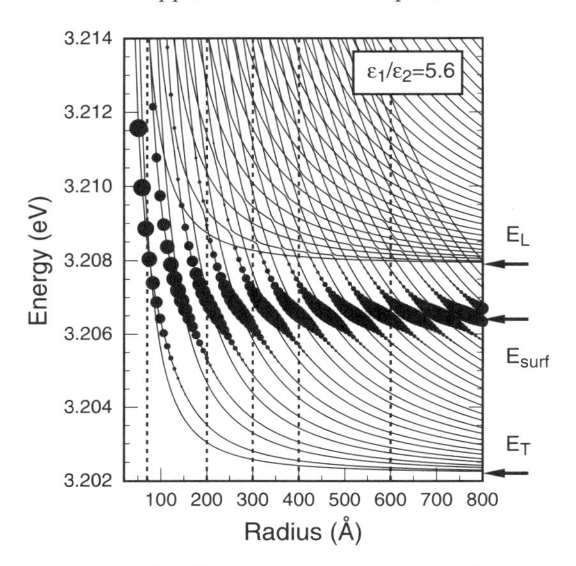

Fig. 4.14. Oscillator strength per unit volume of each confined state at a given R represented by the size of a *solid circle*

The R^3 dependence of the oscillator strength of the lowest exciton state is seen in a still smaller region of R, where the response spectrum consists of a single peak with width given by Fig. 4.15. The straight line corresponds to the standard expression for the radiative width (FWHM) in LWA, i.e., $(4/3)q^3|M_{10}|^2$. The onset of saturation is seen with the increase in R.

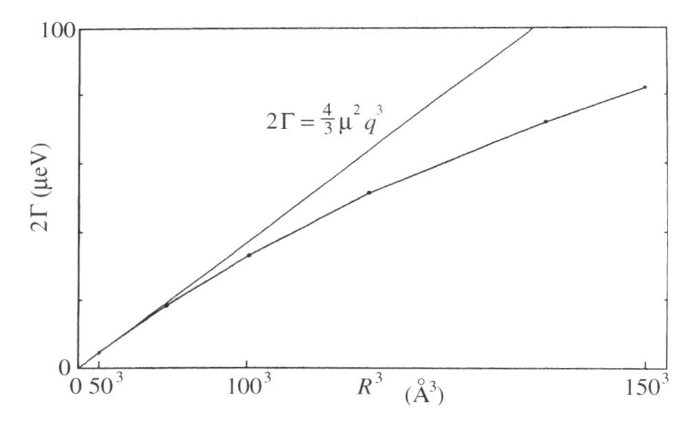

Fig. 4.15. Full width at half maximum (FWHM) of the first peak of the calculated response spectrum for small values of R. The *straight line* shows the radiative width in LWA expected from the golden rule of QED

When calculating the optical response, it is convenient to use the revised form of the nonlocal response theory mentioned in Sect. 3.4.2, where the e–h exchange interaction is included in the generalized radiative correction $\mathcal{A}^{(\mathrm{g})}$ evaluated in terms of the renormalized Green function $\mathbf{G}_{\mathrm{r}}^{(M)}$ for a sphere. The SS modes in this case are obtained by diagonalizing the coefficient matrix of (3.37). The real and imaginary parts of the SS modes are shown in Fig. 4.16a and b for a wide range of R values.

It should be noted that each size-quantized exciton has a characteristic size R, where its interaction with light becomes maximal. Since the radiative shift and width in such a region is rather large, the ω dependence of the radiative correction becomes significant.

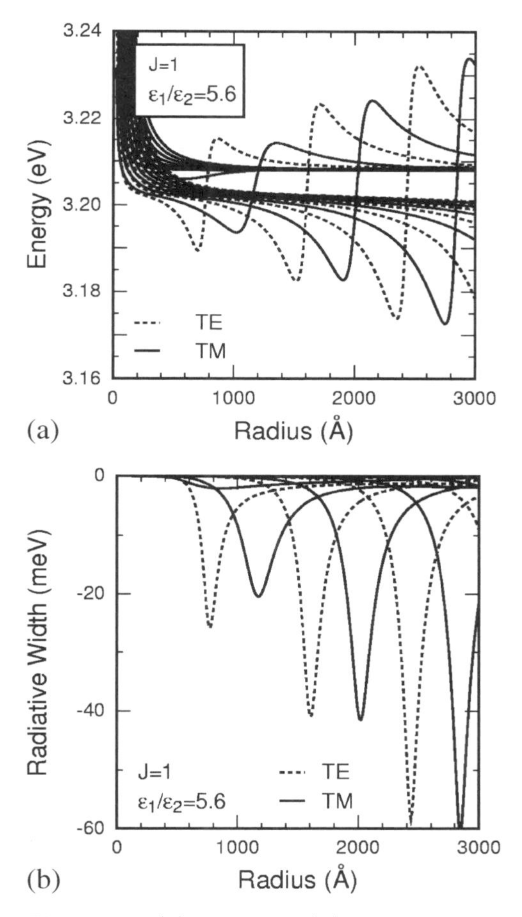

Fig. 4.16. (**a**) Real and (**b**) imaginary parts of the coupled (SS) modes of the sphere with $J = 1$ as a function of R for the same set of parameter values of CuCl

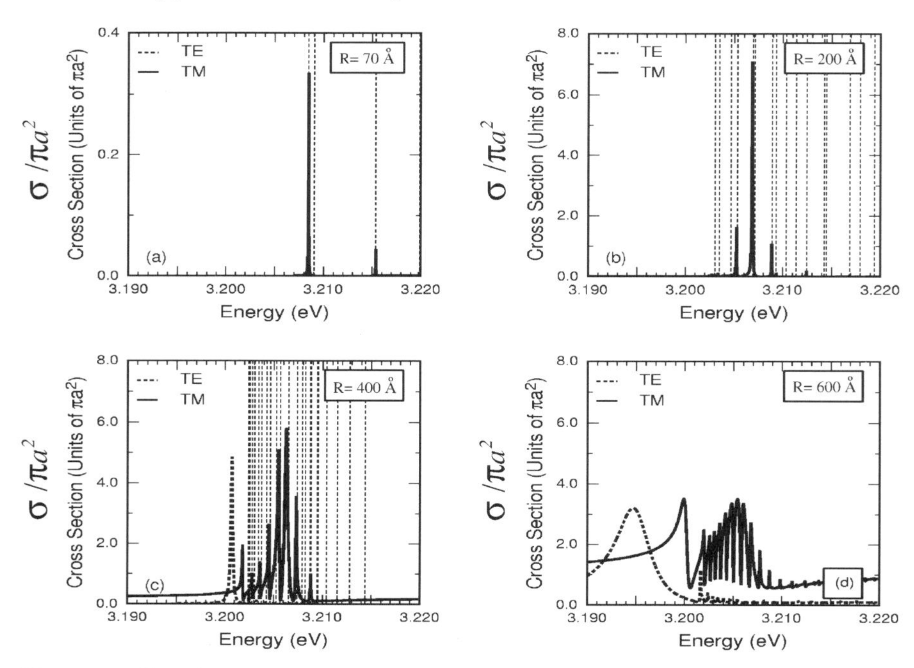

Fig. 4.17. Size dependence of spectral cross-sections for $R = 7, 20, 40, 60$ nm. *Solid* and *dashed curves* are for TM and TE modes, respectively. *Vertical lines* indicate confined exciton levels

Figure 4.17a–d shows the spectra of the cross-section of light for several R values. The figures show the resonant structures due to the TE and TM coupled modes, interfering with the Mie resonance of the sphere. All the resonant structure can be quantitatively analyzed in terms of the size-quantized CM energy, the screened e–h exchange energy, and the radiative shift and width. The near-crossing situation of the lowest LT and T levels, mentioned above, can also be seen in Fig. 4.17a.

Finally, let us consider the relationship between this nonlocal scheme and the macroscopic one due to Ekimov et al. and Ruppin in calculating the response spectrum. As we have used the non-escape boundary condition for the CM motion of the weakly confined exciton in the microscopic model of the nonlocal calculation, which is the microscopic model used to derive the Pekar ABC for a slab or a semi-infinite medium [49], the two results are expected to be essentially the same. According to the numerical test, they actually coincide with each other very accurately in spectral shape and magnitude, if the basis set of the confined exciton in the nonlocal approach is taken large enough with respect to the angular and radial parts of the CM motion [79]. We may thus use the macroscopic method with the Pekar ABC to calculate the response spectrum, unless we require detailed knowledge of the exciton level scheme with the e–h exchange interaction including screening effects, and the T and L character, the radiative shifts and widths, and the interference between the exciton and Mie resonances, etc.

4.1.4 Resonant Bragg Scattering from a Finite Crystal

X-ray diffraction from crystals makes use of the interference effect of the scattered waves from a regular array of atoms. The scattering is described in terms of a dielectric constant with periodic spatial structure, which reflects the periodic array of atoms in a crystal. Under non-resonant conditions, the periodic structure is just the reflection of the charge density in the ground state [27]. In the context of the nonlocal response theory, this is seen directly as the first term on the right-hand side of (2.78a). Under resonant conditions, the remaining terms in the same equation also become important, so that the scattering intensity cannot be ascribed solely to the charge density in the ground state.

In the X-ray region, the radiation–matter interaction is generally weak. But in the lower frequency range, it is much stronger, and the resonant effect is more noticeable. Hence a periodic structure with spacing equal to the wavelength of the resonant EM field would help in realizing a strongly coupled radiation–matter system. Such a system would be useful for manipulating matter by radiation, and vice versa. Photonic crystals with dielectric constants having periodic structures may be regarded as an example of such systems. However, as long as we deal with real dielectric constants, as is usually the case, this is a setup to modulate the mode structure of the EM field, i.e., to manipulate an EM field by matter, a rather one-sided effect.

The other type of strongly coupled radiation–matter systems would be those consisting of resonant matter, which requires a complex dielectric constant. In this case, the $\boldsymbol{p} \cdot \boldsymbol{A}$ term of the radiation–matter interaction in (2.31) plays a more important role than the charge density term discussed above. Because of the resonance, it is essential to consider the self-consistency of the EM field and induced polarization, and this is a more appropriate example of mutually manipulating systems of radiation and matter. In particular, matter resonances are also affected by the EM field, which causes radiative shift and broadening. Since we will discuss photonic crystals later, in Sects. 4.4 and 4.5, we concentrate on the finite regular structure of resonant matter in this section.

Another motivation for this study is the emergence of new types of periodic system. Since the mid-1980s, there have been studies of nuclear Bragg diffraction for crystals of Mössbauer nuclei [80], using a synchrotron radiation source. One of the remarkable findings is the occurrence of the speed-up effect in the radiative decay of excited nuclei [28]. This effect depends sensitively on the Bragg condition. It indicates the importance of geometrical factors in the coupling strength between matter and radiation. Another new example of periodic systems is provided by laser-cooled atoms such as Rb in an optical lattice [81]. These exhibit Bragg diffraction. Although it is a very thin matter system, one can observe the scattered visible light because of the resonant process. It is not a solid in the usual sense, but the mechanism of coherent coupling with the probe ligh t is very similar to that for X-ray diffraction in solids.

Let us begin with a linear chain of N atoms, which we have already considered in Sect. 4.1.1. We reconsider the same model with the question: what happens if we make the lattice constant a_0 larger and larger for a fixed number of atoms? The situation will change qualitatively according to the relative magnitudes of the system size Na_0 and the wavelength λ.

If we start from the case $\lambda \gg Na_0$, the coupling with radiation is monopolized by the single state with the least spatial variation in the induced polarization, i.e., with the smallest of the wave numbers of the envelope function $\{k_n = n\pi/(N+1)a_0,\ (n = 1, 2, 3, \ldots, N)\}$, while the other states have little coupling. (To be precise, some of the dipole active states with k_n for odd n have small but appreciable coupling.)

As the lattice constant a_0 increases, LWA is eventually no longer valid. In such a situation, Fig. 4.2 indicates that some of the states with lower k_n have saturated coupling strength, with saturation value fixed at the single-atom value times $\sim (\lambda/a_0)^1$, as discussed in Sect. 4.1.1. At the same time, a few states with neighboring values of k_n begin to couple more strongly with the radiation. Thus, as we further increase a_0, the saturated value of the enhanced coupling decreases to a value as small as the single-atom value, where $\lambda \approx a_0$. By this time, the mechanism of enhanced coupling via the coherence of size-quantized states is lost.

A new radiative decay channel of excited atomic states emerges when a_0 is large enough to satisfy the Bragg condition $(2a_0/\lambda)\sin\theta$ = integer. If the number of atoms N is not too small, the chain has quasi-1D periodicity, so that the wave number along the chain direction $k_{\|}$ may be considered as a good quantum number for both light and matter systems. Then an excited state of the chain of atoms with quantum number k_n can emit light in the particular directions satisfying the energy and wave vector conservation rules. Because of the finiteness of the chain length Na_0, there is uncertainty in $k_{\|}$, allowing a certain width in the direction of light to be emitted. When a new Bragg condition is satisfied, a new direction of emitted light arises, making an additional contribution to the radiative decay rate of the excited state $|k_n\rangle$.

Figure 4.18 shows $\mathrm{Im}[\mathcal{A}_{\nu 0,0\nu}]$ for $\nu = k_n$ as functions of a_0/λ, calculated for a chain of $N = 100$ hydrogen atoms [82]. The positions of the sharp jumps correspond to the Bragg condition. The condition for the emitted light with jth order Bragg diffraction to be a propagating mode is written as $(2\pi/\lambda)^2 - (k_n + 2\pi j/a_0)^2 \geq 0$, where $j = 0, \pm 1, \pm 2, \ldots$, and λ is the wavelength of resonant light, i.e., $E(k_n) = 2\pi\hbar c/\lambda$. The threshold of a new channel corresponds to a zero of the above inequality, i.e.,

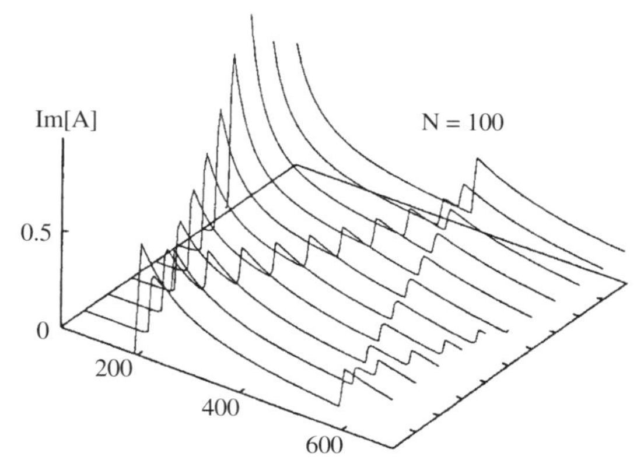

Fig. 4.18. Radiative widths as functions of the lattice interval a_0 for the eigenmodes of a 1D chain of 100 atoms specified by the wave number k'

$$k_n = \pm\frac{E(k_n)}{\hbar c} - \frac{2\pi j}{a_0} , \tag{4.37}$$

which coincides with the positions of the jumps in the figure.

The sudden increase in the radiative width at the threshold involves, in addition to opening up a new decay channel as mentioned above, the effect of enhancing the matrix element of the radiation–matter interaction due to positive interference with the contribution from individual atoms at the Bragg condition. This mechanism plays the main role in the case of a stack of 2D planes discussed later in this section.

For a given incident field, the scattered amplitude is calculated as follows. We first solve (3.5), viz., $\mathbf{S}_{\mathrm{x}}\tilde{\boldsymbol{X}} = \tilde{\boldsymbol{F}}^{(0)}$, where the coefficient matrix is given in the rotating wave approximation as

$$S_{i\xi,j\eta} = (E_{\mathrm{a}} - \hbar\omega)\delta_{ij}\delta_{\xi\eta} + \mathcal{A}_{i\xi0,j0\eta} . \tag{4.38}$$

Here E_{a} is the excitation energy of an atom, (i, j) are the atomic sites, (ξ, η) are the indices of excited states, and the radiative correction is defined as

$$\mathcal{A}_{i\xi0,j0\eta} = \frac{-1}{2\pi^2 c^2}\int \mathrm{d}\boldsymbol{k}\frac{\tilde{\boldsymbol{I}}_{\xi0}(-\boldsymbol{k})\cdot\left[\mathbf{1} - \hat{\mathbf{e}}_3(\boldsymbol{k})\hat{\mathbf{e}}_3(\boldsymbol{k})\right]\cdot\tilde{\boldsymbol{I}}_{0\eta}(\boldsymbol{k})}{k^2 - (q + \mathrm{i}0^+)^2}\mathrm{e}^{\mathrm{i}\boldsymbol{k}\cdot(\boldsymbol{R}_i - \boldsymbol{R}_j)} . \tag{4.39}$$

In this expression, we have used the Fourier coefficient of the current density,

$$\tilde{\boldsymbol{I}}_{\mu\nu}(\boldsymbol{k}) = \int \mathrm{d}\boldsymbol{r}\mathrm{e}^{-\mathrm{i}\boldsymbol{k}\cdot\boldsymbol{r}}\boldsymbol{I}_{\mu\nu}(\boldsymbol{r}) , \tag{4.40}$$

which is common to all the atoms. Because of the large separation between atoms in this model, i.e., $a_0 \sim \lambda$, the states $|\mu\rangle$ and $|\nu\rangle$ in the above expression are only taken within the same atom, and the dipole–dipole interaction in

$\mathbf{S}_x$ is neglected. For an incident field $\boldsymbol{A}_0(\boldsymbol{r}) = \boldsymbol{A}_0\exp(\mathrm{i}\boldsymbol{k}_0\cdot\boldsymbol{r} - \mathrm{i}\omega t)$, the factor $\tilde{\boldsymbol{F}}^{(0)}$ is defined, also in RWA, by

$$F^{(0)}_{i\xi 0}(\omega) = \boldsymbol{A}_0\cdot\tilde{\boldsymbol{I}}_{\xi 0}(-\boldsymbol{k})\exp(\mathrm{i}\boldsymbol{k}_0\cdot\boldsymbol{R}_i) \ . \tag{4.41}$$

The vector potential at a remote point $\boldsymbol{r}$ in the direction $\boldsymbol{k}_\mathrm{s}$ from the sample is

$$\boldsymbol{A}(\boldsymbol{r},\omega) = \boldsymbol{A}_0(\boldsymbol{r},\omega) + \frac{1}{c^2}\sum_i\sum_\xi \frac{\mathrm{e}^{\mathrm{i}k_0 r}}{|\boldsymbol{r}|}X_{i\xi 0}(\omega)\mathrm{e}^{-\mathrm{i}\boldsymbol{k}_\mathrm{s}\cdot\boldsymbol{R}_i}\tilde{\boldsymbol{I}}_{0\xi}(\boldsymbol{k}_\mathrm{s}) \ , \tag{4.42}$$

where $\boldsymbol{k}_\mathrm{s}$ is the wave vector of the scattered plane wave, and we have made the following approximation in the radiation Green function for the far-field condition ($|\boldsymbol{r}| \gg |\boldsymbol{r}'|$):

$$\frac{\mathrm{e}^{\mathrm{i}q|\boldsymbol{r}-\boldsymbol{r}'|}}{|\boldsymbol{r}-\boldsymbol{r}'|} \approx \frac{\mathrm{e}^{\mathrm{i}qr}}{r}\mathrm{e}^{-\mathrm{i}\boldsymbol{k}_\mathrm{s}\cdot\boldsymbol{r}'} \ . \tag{4.43}$$

For an incident $\boldsymbol{k}_0$, the angular distribution of scattered light has sharp peaks at angles satisfying the Bragg conditions for an infinite 1D lattice. As an example, let us consider the case where $\boldsymbol{k}_0$ is parallel to the chain (parallel to the z-axis) and the polar angle of $\boldsymbol{k}_\mathrm{s}$ is θ. Then, from energy conservation, $k_0 = k_\mathrm{s}$, and the Bragg condition, $k_{\mathrm{s}z} = K_0 - 2\pi\mathrm{j}/a_0$, we have

$$\tan\theta = \frac{\sqrt{1-(1-\mathrm{j}\lambda_0/a_0)^2}}{1-\mathrm{j}\lambda_0/a_0} \ , \tag{4.44}$$

where $\lambda_0 = 2\pi/k_0$.

An example of the angular distribution of the scattered intensity calculated from the far-field expression for the vector potential is given in Fig. 4.19 for $a_0/\lambda_0 = 2$, which agrees with (4.44). The effect of the finiteness of the

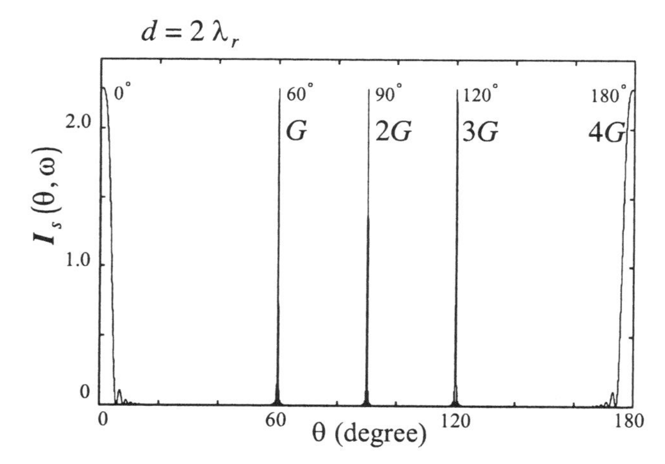

Fig. 4.19. Intensity of scattered light as a function of scattering angle

chain appears in the width and height of each peak and in the zero level of Fig. 4.19. The width and height of the peaks are proportional to $1/\sqrt{N}$ and N^2, respectively, and the zero level consists of a wavy curve on an expanded scale, which gets smaller as N increases [84].

Now let us consider another array of the same atoms as above, i.e., N square lattices in the xy plane stacked in the z direction with spacing $b_0 \sim \lambda$. Each square lattice is infinitely large in the xy plane, and its lattice constant a_0 is assumed to be much smaller than b_0. In this case, we can expect only zeroth order resonant Bragg reflection. In each plane, atomic excitations form Frenkel excitons with lateral wave vector $\boldsymbol{k}_\parallel$. If we allow three-fold (x, y, and z polarized) excited states for each atom, there are three bands of Frenkel excitons as shown in Fig. 4.4. The dipole–dipole interaction between neighboring planes adds small ($k_\parallel$-dependent) dispersion for the wave number in the z direction, $k_{zn} = n\pi/(N+1)b_0$, $(n = 1, 2, 3, \ldots, N)$.

The radiative width of an exciton state with quantum numbers $\{\boldsymbol{k}_\parallel, k_{zn}\}$ is shown in Fig. 4.20 as a function of b_0/λ. The value of $\mathrm{Im}[\mathcal{A}(\omega = E_\mathrm{a}/\hbar)]$ is plotted in the figure [83]. Here again, a resonant enhancement of $\mathrm{Im}[\mathcal{A}]$ is seen at each Bragg condition (4.37). However, there is a remarkable difference between Fig. 4.18 and Fig. 4.20. Indeed, the jump in the latter forms a simple peak, while it is followed by a terrace in the former. This is due to the difference in the enhancement mechanisms. As mentioned in the paragraphs before and after (4.37), the light emission in a new direction at a new Bragg condition is not allowed in the latter case, because of $\boldsymbol{k}_\parallel$-conservation. Therefore, enhancement is solely due to the coupling matrix element at each Bragg condition, which is k_n-dependent, as seen from (4.37). In the former case, the terrace part is assigned to the opening up of a new emission channel.

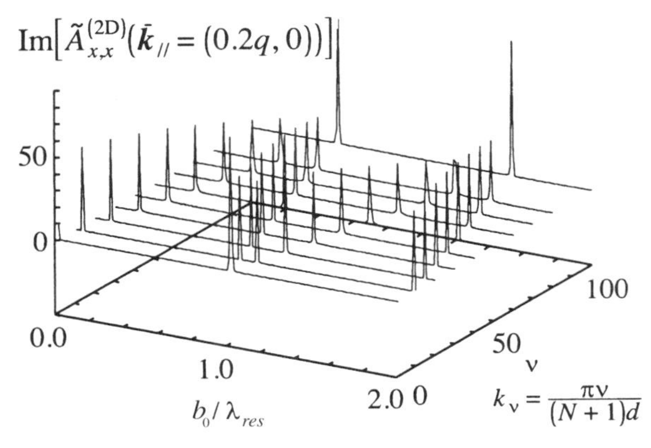

Fig. 4.20. Radiative widths of the excitons in the array of 100 square lattices of atoms as functions of the interlayer spacing in units of resonant wavelength b_0/λ_res. The wave number k_ν specifies the eigenmodes. The lateral component of the wave vector is fixed at $(0, 2q, 0)$, and x-polarized states are considered. The *ordinate* is given in units of the radiative width for a single layer, i.e., 2.6 meV

The magnitude of Im[$\mathcal{A}$] differs enormously in the two cases. Its value in Fig. 4.20 is given in units of the value for a single atomic plane. For $k_{\|} \leq q$, this is enhanced by the factor $(\lambda/a_0)^2$ over the single-atom value, as mentioned in Sect. 4.1.1. Altogether, the enhancement of Im[$\mathcal{A}$] for 100 square lattices is about 10^4 in this example. Noting that the radiative width is an indicator of the radiation–matter coupling strength, we can expect a very strong coupling of this stack of 2D planes for an appropriate geometry.

Regular arrays of resonant matter with Bragg geometry are very interesting if we seek material systems which have optimal coupling with an EM field, as discussed above for the 1D arrays of atoms and atomic planes. In each of these systems, a superradiant (SR) mode arises, which monopolizes the coupling with the light, for small sizes N.

Figure 4.21 shows the radiative widths of the SR modes for the 1D arrays of atoms, atomic chains, and atomic planes, obtained from the reflectivity spectra, as functions of N [84]. The lattice constant b_0 for the 1D arrays is assumed to satisfy a Bragg condition $b_0 = \lambda_{\rm res}/2$. The values at $N = 1$ show the enhancement factor $(\lambda/a_0)^D$ discussed earlier in this section. The figure indicates a further enhancement of the coupling with increasing size N of the 1D arrays. This means that there are two enhancement mechanisms, i.e., a regular arrangement of atoms with the normal lattice constant $(\lambda/a_0 \gg 1)$ in a chain or a plane, and the Bragg arrangement.

Figure 4.22 shows the corresponding reflectivity spectra for various N values for the arrays of atomic planes. For small N, the spectrum has a Lorentzian form with width $N\Gamma_0$, where Γ_0 is the width of a single plane. As N becomes larger (Fig. 4.22b), the spectral shape becomes distorted from Lorentzian, and converges to a top-hat shape, indicating the formation of

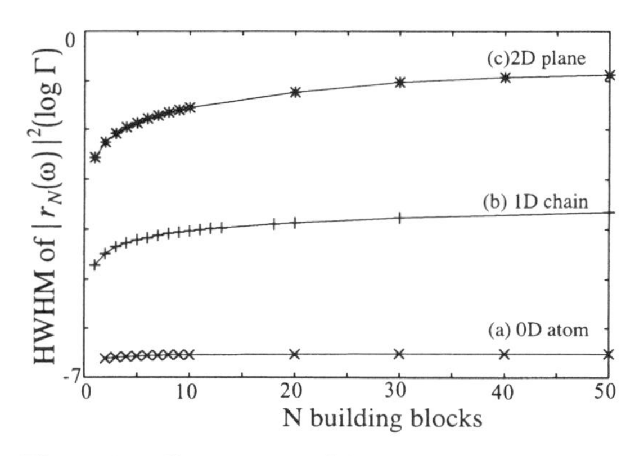

Fig. 4.21. Comparison of the size dependence of the radiative widths for 1D arrays of N atoms, N atomic chains, and N atomic planes satisfying the Bragg condition. N-linear (N-sublinear) growth is seen for the atomic planes (chains) for small N values

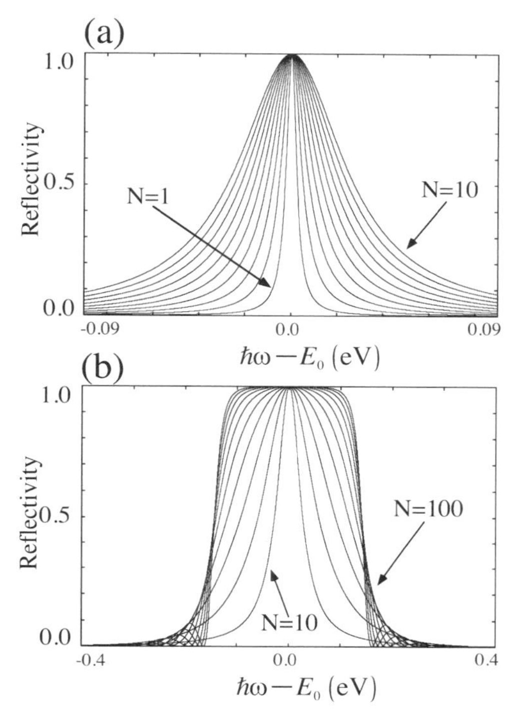

Fig. 4.22. Reflectivity spectrum for a 1D array of N atomic planes. (**a**) $N = 1, 2, \ldots, 10$, (**b**) $N = 10, 20, \ldots, 100$

total reflection range. This is reasonable, in view of the fact that the arrays with large N can be regarded as a photonic crystal. The total reflection range corresponds to its photonic gap [84].

This evolution from the SR mode to a photonic gap demonstrates the peculiar nature of this system amongst materials which have a strong coupling with light. For small N, it has an SR mode with an intense field around it, but in the photonic crystal regime for large N, it has a gap repelling the EM field completely from the crystal. One might ask what becomes of the SR mode with increasing N. The fate of the SR mode can be understood from the ω dependence of the radiative correction [85]. Among the various excited states of the array, which can be distinguished by the wave number k_z in the direction of the stack and appropriately quantized for finite N, the one with $k_z = \pi/b_0$ constitutes the SR mode, and all the others have almost no coupling with light for small N systems. If we consider only the SR mode in calculating the response spectrum, which is described by the radiative correction of the SR mode $\mathcal{A}_{\rm SR}(\omega)$ alone, the cases of small N are well reproduced, but the deviation from the exact spectrum which takes all

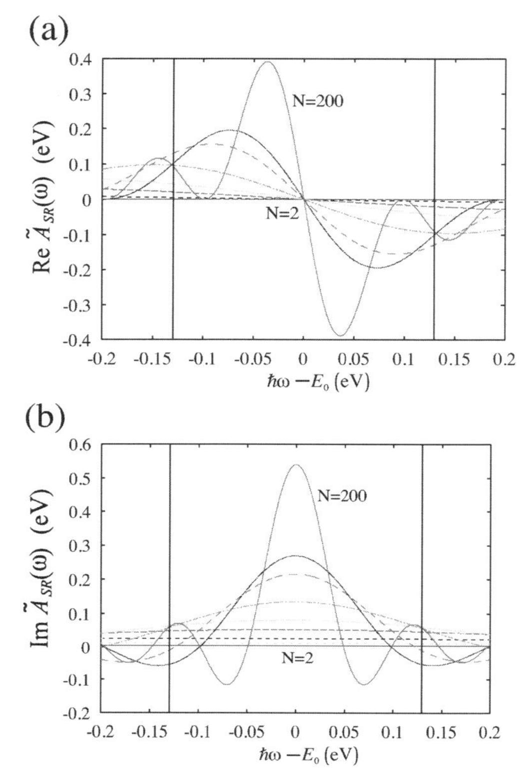

Fig. 4.23. (**a**) Real and (**b**) imaginary parts of the radiative correction of the SR mode in the array of N atomic planes as functions of ω for $N = 2, 5, 10, 20, 30, 40, 50, 80, 100, 200$. The *vertical solid lines* indicate the positions of the total reflection range for the $N = \infty$ system

the excited states into account becomes more serious for larger N. The reason for this becomes clear if we calculate $\mathcal{A}_{\mathrm{SR}}(\omega)$ for various N, as in Fig. 4.23.

It is evident from this figure that $\mathcal{A}_{\mathrm{SR}}(\omega = E_0/\hbar)$ is only a good approximation for small N. Although $\mathcal{A}_{\mathrm{SR}}(\omega)$ describes the spectrum for any N, its interpretation as the radiative shift and width becomes invalid for large N. In this sense, the SR mode is lost in the system with large N.

The spectral evolution from the SR mode to the photonic crystal can also be calculated using the transfer matrix method [86]. The evolution can be ascribed to the ω dependence of the transfer matrix elements, but it does not explain the physics very well. The physical interpretation is most appropriately made within the nonlocal framework, as mentioned above, in terms of the growing importance with N of the ω dependence of the radiative correction, which is defined from first principles as an ω-dependent quantity, and incorporated naturally in a self-consistent scheme of optical response.

In connection with the photonic crystal regime of this system, there is a new feature in the photonic bands, i.e., a propagating mode in the photonic gap. Details will be given in Sect. 4.5.

4.2 Cavity Mode Coupled with a Resonant Level

In this section, we consider a somewhat specialized situation concerning the EM field, i.e., the case of cavity mode formation. In free space, the eigenmodes can be expressed by plane waves, which have a continuous spectrum with density of states proportional to ω^2. Confining the EM field in a cavity, we can prepare various mode structures including discrete spectra, which can change the radiation–matter interaction in a drastic way.

For a given mode structure of the EM field, the expression for the radiative rate for the transition between levels 0 and 1 is given by the golden rule in QED [44]:

$$W = \frac{2\pi}{\hbar} \sum_{\lambda} |g_\lambda|^2 \delta(\hbar\omega_\lambda - E_{10}) , \tag{4.45}$$

where λ specifies the mode of the EM field, and g_λ is the coupling strength between the λ-mode photon and the relevant transition with resonant energy E_{10}.

This expression may be regarded as the density of states of light weighted by the coupling strength. Therefore, a change in the mode structure is sensitively reflected in the value of the radiative rate. For example, if the density of states is zero in a certain frequency range, a matter state with excitation energy in this range cannot emit light. The change is also accompanied by the characteristic spatial structure of each mode, which leads to a specific dependence of the radiation–matter coupling on geometrical factors such as position, size, shape, etc. of a sample in (or near) the cavity. Studies of these aspects of quantum optics are called cavity QED, which is now very popular [24].

For theoretical discussions, we usually need well-defined cavity modes, the number of which is desired to be as small as possible for a clearer mathematical treatment. It is often assumed that each mode has damping characterized by appropriate parameters. The origin of the damping could be the incompleteness of the cavity, or the lossy dielectric properties of the material used for the cavity. The simple-minded introduction of a damping mechanism in cavity QED destroys the commutator relationship of various operators in the course of time. The recipe used to recover it involves introducing the fluctuation force satisfying the fluctuation–dissipation theorem [87] or Lindblad operators [88].

The incompleteness of the cavity means that all external modes (plane waves, essentially, with continuous spectrum) can generally have finite amplitudes in the cavity with different weights. The energy range with locally maximum weight of the amplitude in the cavity corresponds to a cavity mode. The mode calculation in this case requires all the details of the cavity, i.e., its shape and size, and the material composing it. Actually, this is a more realistic situation than the case of an idealized lossless cavity. It is even necessary to have finite Q-factors to observe the radiation–matter coupling as a signal of the EM field outside the cavity because, if the confinement of the field is 100%, no light comes out of the cavity at the corresponding frequency.

The subject of cavity effects on the optical properties of matter extends beyond the area usually referred to as cavity QED, to include both fundamental science and device applications. Even the definition of a cavity might well be generalized to any (part of) material which modifies the mode structure of an EM field. In this sense, any material is indeed more or less a cavity, if the modified part of the EM field affects the radiation–matter coupling in an appreciable way. From this viewpoint, we consider the following systems in this section in terms of the microscopic nonlocal theory:

- an atom coupled with a slab,
- an atom coupled with WG modes,
- a QW exciton in a microcavity.

4.2.1 Atom Coupled with a Slab

In general, the presence of a dielectric medium in vacuum changes the mode structure of the EM field. The case of a uniform slab with isotropic dielectric constant $\epsilon_b (> 1)$ is one of the simplest examples. The standard way to calculate the influence of the slab on the radiative decay rate of an atom inside or outside the slab is to use the golden rule (4.45). The complete set of the EM field for the slab in vacuum can be chosen as the plane waves with components parallel and perpendicular to the slab (Fig. 4.24).

Due to the law of refraction, the plane waves in the slab have evanescent tails outside the slab (wave guide modes), when the incident angle to the surface exceeds the critical angle $\arcsin(1/\sqrt{\epsilon_b})$. Otherwise, the travelling plane waves inside the slab are connected to those outside the slab by means of the MBC. The spontaneous decay of a matter excitation leaves the matter in the ground state and a photon in one of the modes of the complete set satisfying the energy conservation rule. We may divide the contribution into that of the travelling plane waves and that of the wave guide modes. Because of confinement in the slab, the latter generally have large amplitudes inside and just outside the slab, which leads to a large contribution from these modes to the decay rate of an atom lying in this region.

The amplitudes of the travelling plane waves contain the interference of incident and reflected plane waves on one side of the slab. This gives an oscillating feature to the amplitudes of this mode as a function of the distance h_0 from the surface of the slab. Because of this, the contribution of these modes to the decay rate of the atom gives an oscillating component as a

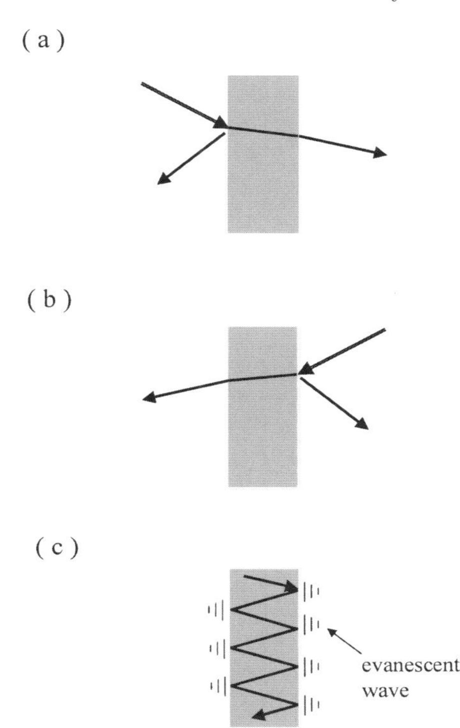

Fig. 4.24a–c. A complete set for the EM field within a dielectric slab, consisting of plane waves incident from left and right infinity and waveguide modes. All of them are specified by the lateral wave vector $\boldsymbol{k}_{\|}$

function of h_0. For large values of h_0, the effect of the slab should vanish, and hence the limiting value of the decay rate for $h_0 \to \infty$ is that for an atom in vacuum. Such behavior is seen in the example shown later.

The other method for calculating the decay rate is the one based on the nonlocal treatment of optical response. We only need to calculate the imaginary part of the radiative correction $\mathcal{A}^{(g)}_{10,01}$ in (3.38), where the radiation Green function contains the effect of ϵ_b of the slab. The analytical form of the renormalized radiation Green function $\mathbf{G}_r^{(M)}$ is known [26], and this allows a straightforward calculation of the decay rate.

Figure 4.25a, b and c shows $\mathrm{Im}[\mathcal{A}^{(g)}_{10,01}]$ as a function of atom position h_0 and polarization of the induced dipole moment. The result of the QED calculation turns out to be exactly the same as that of the semiclassical nonlocal theory. In the figures, the separate contributions of waveguide (trapped) modes and travelling plane waves in the QED calculation are also shown.

It should be noted that the h_0 dependence across the slab surface is continuous for polarization parallel to the surface and discontinuous for perpendicular polarization. This is simply the reflection of the MBC for the EM field at the boundary, which is used to construct the complete set or the radiation Green function. The wavelength of the oscillating component outside

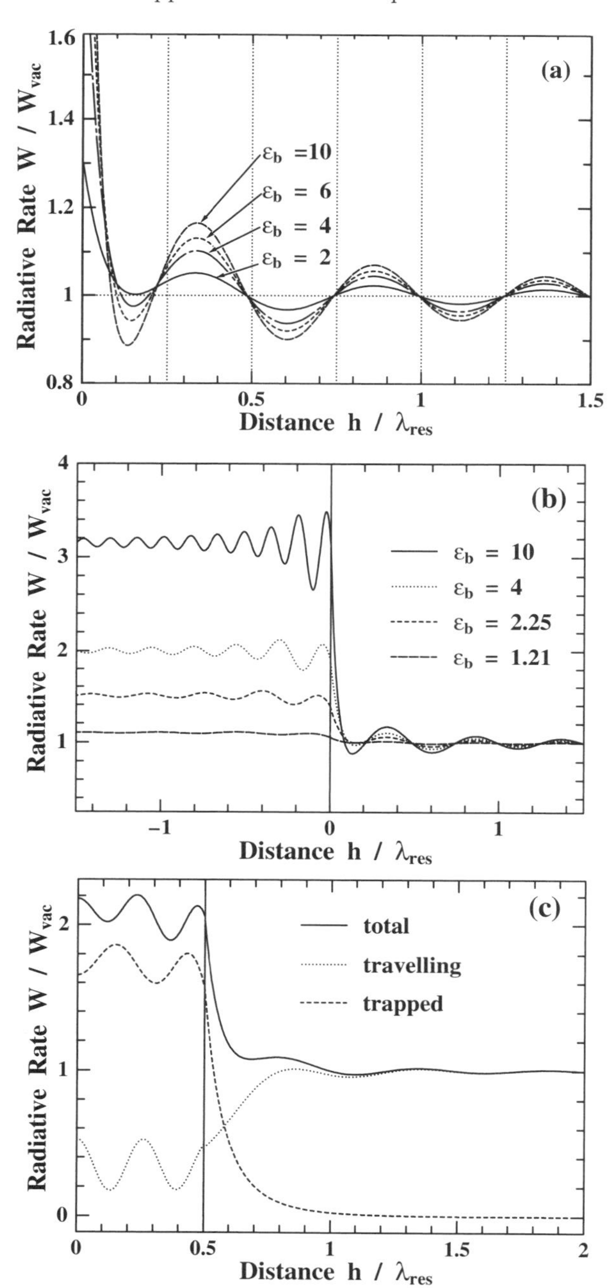

Fig. 4.25. Radiative decay rate of an atom (**a**) outside, (**b**) inside and outside a semi-infinite dielectric, and (**c**) inside and outside a slab, as a function of its position for the case of polarization parallel to the surface. (**a**) is the result of nonlocal calculations, whilst (**b**) and (**c**) come from the QED golden rule. *Vertical lines* in (**b**) and (**c**) indicate the position of the surface [91]

the slab is close to that of the resonant EM wave. The large contribution of the waveguide modes for an atom in the dielectric can also be seen in the figures. For a slab thickness comparable to the resonant wavelength, the main part of the emitted light goes into the waveguide modes.

4.2.2 Atom Coupled with WG Modes

As mentioned in Sect. 3.2, a WG mode is a self-sustaining (SS) mode of a dielectric sphere. Within the framework of macroscopic response, its complex eigenfrequency is obtained from (3.10) as $\det|\mathbf{S}'| = 0$ applied to a sphere. This is equivalent to the condition that the scattering amplitude diverges. The real and imaginary parts of the complex frequencies correspond to the resonant frequencies and widths, respectively, of the eigenmodes. Although the condition $\det|\mathbf{S}'| = 0$ gives solutions for any size of the radius R, the term 'WG mode' is used when the radius is larger than the wavelength λ, so that the picture of multiple internal reflection of light with little leakage is more appropriate. On the other hand, the SS modes for smaller radii ($R \sim \lambda$) are usually called Mie resonances. As a cavity mode, a WG mode can have a very large Q-factor. This provides an a ppropriate situation for understanding the effect of the coupling with another resonance.

If the local dielectric constant ϵ_b of the sphere is real, the width of the modes is purely radiative. The radiation–matter interaction contributing to the radiative width is also responsible for the formation of new coupled modes in the presence of another resonant excitation. On the other hand, for the complex value of ϵ_b, the imaginary part of which originates from non-radiative scattering, there arise WG modes containing both radiative and non-radiative widths. An example of this case is a dielectric sphere doped with dye molecules. In this case, ϵ_b is ω dependent, reflecting the absorption spectrum of the dye molecules. WG modes are determined as the complex roots of $\det|\mathbf{S}'| = 0$ as before. Their widths (imaginary parts of the roots) are increased in comparison with the undoped case, reflecting the form of the absorption spectrum of the dye molecules.

An example is shown in Fig. 4.26, where the calculated widths of WG modes in the presence and absence of dye molecules are compared [89]. By doping dye molecules, the width of each WG mode is increased in an ω-dependent way, reflecting the absorption spectrum of the dye. The behavior of the WG modes in a doped dielectric sphere can be seen in luminescence spectra [90]. This example may be regarded as the case of sharp WG modes in an undoped sphere coupled with the broad resonance of dye molecules whose width has non-radiative origin.

Another case of interest is the coupling of a sharp WG mode with a sharp atomic excitation [91], where the atom lies inside or outside the sphere. Because of the high Q-factors of the two resonances, the effect of the coupling will be seen as new modes with a shift in the resonance energies and a change in their widths. In this case, it is more convenient to calculate the coupled

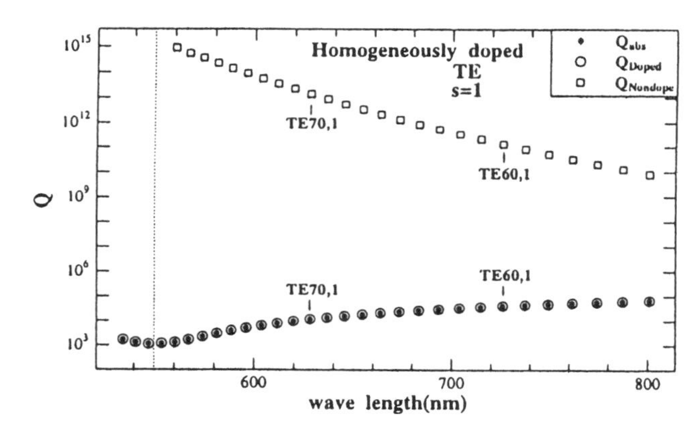

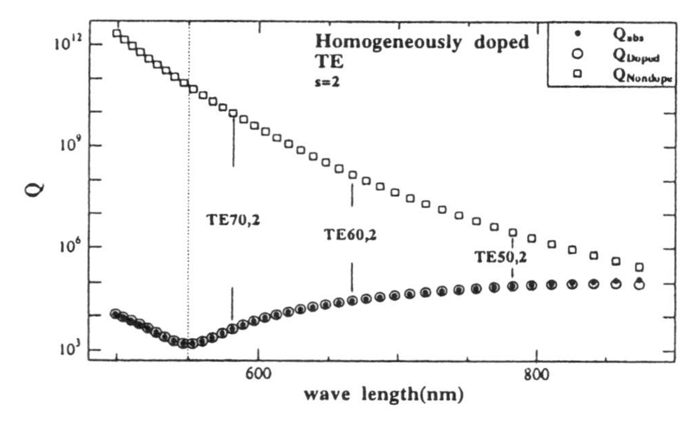

Fig. 4.26. Q-factors of WG modes (TE, $s = 1, 2$) in a dielectric sphere ($R = 10\,\mu$m, $\epsilon_\mathrm{b} = 1.578^2$) homogeneously doped with dye molecules. The *dotted vertical line* shows the position of the absorption peak of the dye. *Empty squares* represent the case for undoped spheres

modes as the complex roots of det $|\mathbf{S}| = 0$ in the nonlocal formulation. According to the reformulation in Sect. 3.4.2 to renormalize the effect of ϵ_b of the sphere, the condition for the existence of SS modes is the vanishing of the determinant of the coefficient matrix in (3.37). This includes the radiative correction term $\mathcal{A}^{(\mathrm{g})}_{10,01}$ evaluated in terms of the renormalized radiation Green function $\mathbf{G}_\mathrm{r}^{(\mathrm{M})}$. If the atom is described as a two-level system, equations (3.37) reduce to a single equation ($\nu = 1$), and the condition for SS modes is

$$E'_{10} - \hbar\omega + \mathcal{A}^{(\mathrm{g})}_{10,01} = 0\,. \tag{4.46}$$

In this formulation, the (complex) eigenfrequencies of the WG modes of the isolated sphere are included as the poles of the renormalized radiation

Green function. Therefore, the above equation has roots corresponding to the coupled modes between the atomic excitation and WG modes. Once again, it should be stressed that the formulation in terms of the renormalized radiation Green function $\mathbf{G}_{\mathrm{r}}^{(\mathrm{M})}$ describes not only the simple cases like (4.45), but also more general situations where new coupled modes emerge as a result of the interaction between matter resonance and cavity modes. The golden rule (the second order perturbation) describes the effect of the EM field on the matter, but not the back-reaction on the field.

The analytical form of the renormalized radiation Green function is known for a single sphere [26], and this facilitates numerical evaluation of the roots of the above equation. In doing so, it is useful to take symmetry into consideration. Since each sphere and atom has spherical symmetry, there exists an axial symmetry around the axis connecting the atom and the center of the sphere. Therefore, when considering the atom–radiation interaction, we have conservation of the component (M) of angular momentum along the axis. In particular, interaction is allowed only between the atomic excitation and WG modes with the same value of this projection. The dipole-active excitation of an atom has angular momentum $J_{\mathrm{a}} = 1$, and the WG modes are specified by quantum numbers $J_{\mathrm{WG}}, M_{\mathrm{WG}}$, and s (order number), together with indices TE and TM. Coupled modes can thus be expected for $M = 1, 0, -1$, and WG modes with other M values remain intact.

Figure 4.27 shows the spectrum of scattered light intensity from a dielectric sphere of radius $R = 5$ µm, $\epsilon_{\mathrm{b}} = 5$. The resonant structure corresponds to various WG modes labelled by TM/TE J, and s. Sharp peaks are WG modes with order number $s = 1$. When an atom with excitation energy close to a sharp WG mode is put in the neighborhood of the sphere, coupled modes emerge with the peak positions and widths shown in Fig. 4.28 as functions of the detuning energy of the atomic excitation with respect to the particular

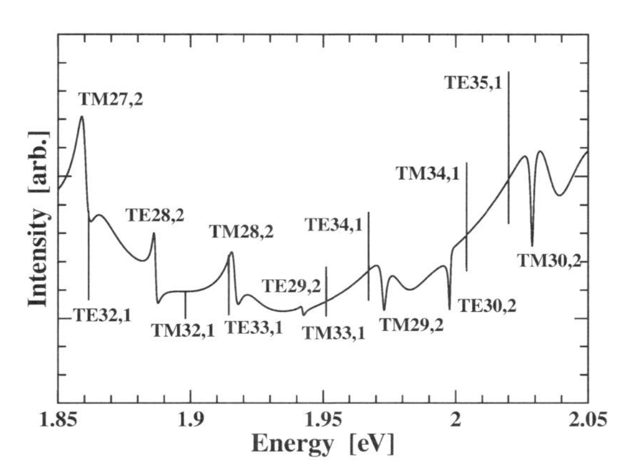

Fig. 4.27. Scattered light intensity from a dielectric sphere with the assignment of WG modes for order number $s = 1, 2$ [91]

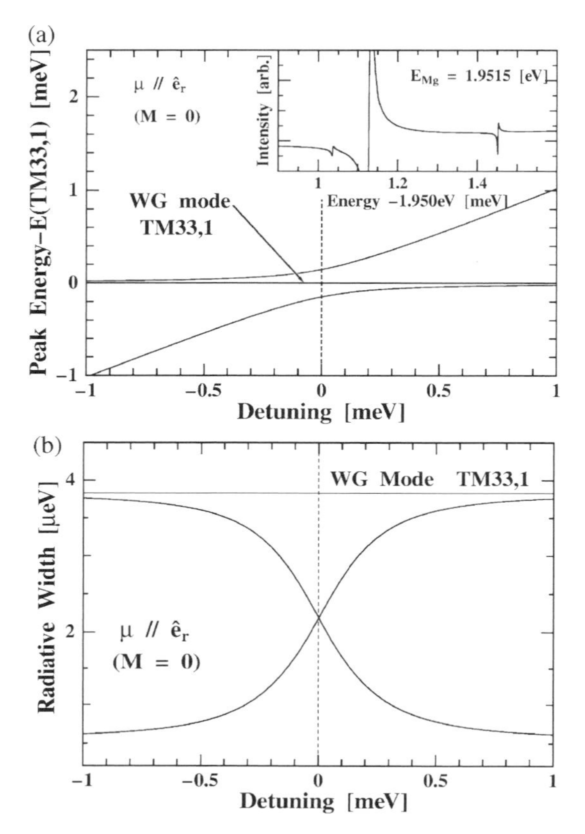

Fig. 4.28. (**a**) Energies of the coupled modes as functions of atom excitation energy measured from the WG mode energy (1.9511 eV) for TM33,1. The dipole moment of the doped 'atom' is assumed to be 50 debye. The *inset* is the spectrum of scattered light intensity for a detuning of +0.4 meV. (**b**) Radiative widths of coupled modes as functions of the detuning. The exchange of the width occurs almost solely between the WG mode and the 'atom' [91]

WG mode with $\hbar\omega(\mathrm{TE33}, 1) = 1.9511$ eV. The inset shows an example of the scattered spectrum for the atomic excitation energy $E_\mathrm{A} = 1.9515$ eV, where the central peak represents the uncoupled components of the WG modes.

4.2.3 Quantum Well Excitons in a Microcavity

Excitons in a quantum well (QW) usually belong to the strong confinement regime, i.e., the confinement potential affects the relative motion as well as the CM motion of the bulk exciton states. The appropriate picture therefore treats e–h pair states of individually confined particles with certain modifications due to the Coulomb attraction.

The size of the confinement, i.e., the thickness of the QW, is comparable to or smaller than the radius of the bulk exciton. In such a situation, the size enhancement of the oscillator strength works in a different direction compared with the weak confinement regime. Indeed, the narrower the QW, the larger the exciton oscillator strength. This is just the reverse tendency of the size enhancement of excitons in the weak confinement regime, as mentioned in Sect. 4.1.1. (The oscillator strength must be compared in common units, because various definitions are used in the literature, such as the oscillator strength per unit cell volume, unit thickness, or unit area. For our present purposes, the oscillator strength per quantized state will be most appropriate, because we then have the same definition as in other parts of this book.)

The mechanism for size enhancement is rather simple in the case of a QW exciton. As the QW thickness decreases, the e–h relative motion changes from 3D to 2D type. In the 2D exciton [92], the Coulomb attraction acts more strongly than in the 3D case, and the exciton binding energy and oscillator strength become larger [93]. In the weak confinement regime, on the other hand, size enhancement of the transition dipole moment arises from the integral over the CM coordinate of the e–h pair, which is proportional to the sample volume in LWA for the state with the smoothest envelope function.

Although the enhancement of the oscillator strength of a QW exciton cannot exceed the 2D limiting value [92,94], the coupling with the radiation field can be further enhanced by putting the QW in a microcavity (MC). For a given frequency range, a cavity can be constructed from distributed Bragg reflectors (DBR), consisting of a (finite) periodic array of dielectrics. Such an array can act as a mirror for a particular frequency range, where the constituent dielectrics are transparent. This is a special case of a photonic crystal, and the DBR is a device making use of a photonic band gap. Some theoretical features of photonic crystals will be discussed in Sects. 4.4 and 4.5.

For an infinite array of two alternating dielectrics (dielectric constants ϵ_a, ϵ_b, thickness $= L_a, L_b$), the lowest gap is expected around the frequency $c\pi/\bar{n}L$, where $L = L_a + L_b$, and $\bar{n}$ is the averaged refractive index $\bar{n} = (n_a L_a + n_b L_b)/(L_a + L_b)$. This is the frequency at which the phase increment of the light for passage over one unit structure in the normal direction is $\pi = (\omega/c)(n_a L_a + n_b L_b)$, which is the Bragg condition for normal incidence. For off-normal incidence with incidence angles θ_a and θ_b in the respective layers (with the law of refraction $n_a \sin\theta_a = n_b \sin\theta_b$), the Bragg condition is

$$\frac{\omega}{c}\left(\frac{n_a L_a}{\cos\theta_a} + \frac{n_b L_b}{\cos\theta_b}\right) = \pi \,. \tag{4.47}$$

For a given structure, the frequency satisfying this relation changes with the incidence angle. Since light of this frequency is mirror reflected by the DBR, the region sandwiched between a pair of symmetrically arranged DBRs becomes a cavity for this frequency. The cavity mode frequency changes with the angle of the propagating direction according to the above equation. For a finite number of double layers in a DBR, the mirror reflection is not 100%, and the Q-factor of the cavity is therefore finite. In this way, we can design the mode frequency, the Q-factor of the microcavity, and the position of the QW with respect to the DBRs, so that we can control the exciton–radiation interaction. By adjusting the resonant energies of the exciton and cavity mode, one can, for example, enhance light emission from the exciton in a selected angle from the surface normal [95]. Such a microcavity with a QW is now available by MBE fabrication. It is used for fundamental and applicational studies of the remarkable features of an exciton–radiation interaction in a MC [96,97].

In this system, if the 2D structure is perfect, both matter and radiation field have translational symmetry along the lateral direction of the QW. Thus the 2D wave vector $\boldsymbol{k}_{\parallel}$ is a good quantum number. The radiation–matter interaction occurs only between the QW exciton and the EM field with the same $\boldsymbol{k}_{\parallel}$.

For each $\boldsymbol{k}_{\parallel}$, the interaction of a QW exciton with a cavity mode of infinitely large Q-factor constitutes a two-level system coupled with a single cavity mode, which can be handled by the well-known Jaynes–Cummings model in cavity QED [98]. Denoting the two levels of the matter by $\{\mathrm{g}, \mathrm{e}\}$ and the photon number by n, we have the radiation–matter interaction between $|\mathrm{g}, n+1\rangle$ and $|\mathrm{e}, n\rangle$, which causes the Rabi splitting [99]

$$E_{\mathrm{Rabi},n} = \sqrt{\hbar^2(\Omega_{\mathrm{a}} - \Omega_0)^2 + 4|C|^2(n+1)} \,, \tag{4.48}$$

where Ω_{a} $[= (E_{\mathrm{e}} - E_{\mathrm{g}})/\hbar]$ and Ω_0 are the resonant frequencies of the matter and cavity mode, respectively. The factor C, which is the coupling matrix element of the radiation–matter interaction, reflects the amplitude of the cavity mode in the two-level system (QW). Hence, the QW at the node (anti-node) of the cavity mode has the minimum (maximum) coupling.

The case $n = 0$ for zero detuning ($\Omega_{\mathrm{a}} = \Omega_0$) gives the vacuum Rabi splitting. The same amount of splitting is obtained from the condition for non-trivial solution in the linear response of the semiclassical treatment applied to the same model. This is reasonable, because the condition for the non-trivial solution gives the eigenmodes (SS modes) of the coupled system.

The semiclassical treatment can be straightforwardly extended to the case of finite Q-factor for the cavity mode, which corresponds to finite DBR thickness. For a given model of the DBRs and QW, we simply set up the equations (3.10) for the components of the internal electric fields and the condition for the SS modes, i.e., the condition $\det|\mathbf{S}'| = 0$ for the non-trivial solution gives the solution of the coupled modes, not only in the absence but also in the presence of a finite width of cavity mode.

The condition for zero detuning can be satisfied for a particular value of $\boldsymbol{k}_{\parallel}$. Hence, while the cavity mode frequency changes with $\boldsymbol{k}_{\parallel}$ (or θ) according to (4.47), the energy of the QW exciton changes little in the range of $\boldsymbol{k}_{\parallel}$ corresponding to the wavelength of the cavity mode. If the exciton energy is higher than the cavity mode at $k_{\parallel} = 0$, the zero detuning condition will be satisfied at a finite value of $k_{\parallel}$.

Generally, the decay of a QW exciton will lead to radiation in a direction with the same $\boldsymbol{k}_{\parallel}$, and this determines the radiative decay width of the exciton. However, in the neighborhood of zero detuning, new coupled modes arise with different resonance energies and widths from those of the exciton and the cavity mode. This is another example of SS modes for a coupled radiation–matter system.

4.2.4 Green Function for the Cavity Polariton

Here we present a new type of radiation Green function which describes the propagation of the EM field inside and outside the cavity polariton system consisting of a QW in a DBR cavity. The purpose is to provide a tool to describe not only the cavity polariton with some leakage to the vacuum, but also the vacuum EM field disturbed by the cavity with a QW, taking cavity structure into account.

This Green function will be useful for calculating the signals due to nonlinear current density [100, 101], phonon-induced current density [102, 103], etc. The optical processes in cavity polariton systems are usually analyzed in terms of the coupling of QW excitons and cavity mode photons. Cavity modes are thought to be well defined, but there is nevertheless an implicit assumption about leakage of the cavity modes into the vacuum and vacuum modes into the cavity. Only in the presence of this leakage can one excite cavity modes via an incident field from outside or detect signals from the QW outside the cavity. The intensity and width of the signals can be quantitatively discussed only by taking this leakage into account, and this will be necessary if we are to analyze the parametric amplification of the modes not lying on the cavity polariton branches [101], i.e., if we wish to describe the leakage, not via phenomenological parameters, but according to the microscopic structure of the cavity.

The Green function to be discussed here solves the 1D Maxwell equation

$$\left(-\frac{\mathrm{d}^2}{\mathrm{d}z^2} - q^2 + \boldsymbol{k}_{\parallel}^2\right) A(z, \boldsymbol{k}_{\parallel}, \omega) = \frac{4\pi}{c} j(z, \boldsymbol{k}_{\parallel}, \omega) , \tag{4.49}$$

where $q = \omega/c$. Here both matter and EM field are assumed to have 2D translational symmetry characterized by a lateral wave vector $\boldsymbol{k}_{\parallel}$, and $A(z, \boldsymbol{k}_{\parallel}, \omega)$ is the lateral component of the vector potential. We decompose (the lateral component of) the current density into three parts

$$j(z) = j_{\mathrm{b}}(z) + j_{\mathrm{QW}}^{(1)}(z) + j'(z) , \tag{4.50}$$

where j_{b} and $j_{\mathrm{QW}}^{(1)}$ describe the current densities due to the cavity and the linear part of the QW exciton, respectively, and j' represents any other components of the current density, such as nonlinear current density, phonon-induced current density, and so on. We assume that the former two components are described in terms of the local and nonlocal linear susceptibilities, respectively, by

$$j_{\mathrm{b}}(z) = \frac{\omega^2}{c} \chi_{\mathrm{b}}(z)\ A(k_{\|}, z) , \tag{4.51}$$

$$j_{\mathrm{QW}}^{(1)}(z) = \frac{J_{0\nu}(k_{\|}, z)}{cs[E_{\mathrm{QW}}(k_{\|}) - \hbar\omega - \mathrm{i}\gamma]} \int \mathrm{d}z' J_{\nu 0}(-k_{\|}, z') A(k_{\|}, z') , \tag{4.52}$$

where s is the area of the lateral unit cell. The background susceptibility $\chi_{\mathrm{b}}(z)$ represents the cavity structure and the background part of the QW, and is described in terms of a multiple step function of z. As the resonant term, we consider only one branch of the QW exciton and all the others are assumed to be included in the χ_{b} of the QW region. The Fourier component of the matrix element of the current density operator for the exciton transition is

$$J_{\mu\nu}(\boldsymbol{k}_{\|}, z) = \int \mathrm{d}\boldsymbol{r}_{\|} \exp(-\mathrm{i}\boldsymbol{k}_{\|}\cdot\boldsymbol{r}_{\|}) J_{\mu\nu}(\boldsymbol{r}_{\|}, z) , \tag{4.53}$$

where $J_{\mu\nu}(\boldsymbol{r})$ represents the unit cell structure of the exciton-induced current density

$$I_{\mu\nu}(\boldsymbol{r}) = \frac{1}{\sqrt{N_{2D}}} \sum_{\boldsymbol{R}_{\|}} \exp(-\mathrm{i}\boldsymbol{k}_{\|}\cdot\boldsymbol{R}_{\|}) J_{\mu\nu}(\boldsymbol{r}_{\|} - \boldsymbol{R}_{\|}, z) , \tag{4.54}$$

with N_{2D} being the number of lateral lattice sites. The Green function of the cavity $g(z, z', k_{\|}, \omega)$ is defined by

$$\left[\frac{\mathrm{d}^2}{\mathrm{d}z^2} + q^2 \epsilon_{\mathrm{b}}(z)\right] g(z, z') = -4\pi\delta(z - z') , \tag{4.55}$$

where

$$\epsilon_{\mathrm{b}}(z) = 1 + 4\pi\chi_{\mathrm{b}}(z) - \frac{k_{\|}^2}{q^2} . \tag{4.56}$$

The solution of (4.55) can be obtained analytically for an arbitrary multilayer structure [26]. In the following, we assume that it is known for any ω and $k_{\|}$.

We now define the Green function of the cavity polariton $G^{(\mathrm{cp})}(z, z', k_{\|}, \omega)$ by

$$\left[\frac{\mathrm{d}^2}{\mathrm{d}z^2} + q^2 \epsilon_{\mathrm{b}}(z)\right] G^{(\mathrm{cp})}(z, z') + \frac{4\pi}{sc} \frac{J_{0\nu}(z)}{[E_{\mathrm{QW}}(k_{\|}) - \hbar\omega - \mathrm{i}\gamma]} H_{\nu 0}(z') = -4\pi\delta(z - z') , \tag{4.57}$$

where

$$H_{\nu 0}(z') = \frac{1}{c} \int \mathrm{d}z'' J_{\nu 0}(z'') G^{(\mathrm{cp})}(z'', z') . \tag{4.58}$$

In terms of this Green function, the solution of the Maxwell equation (4.49) is

$$A(z) = A_0(z) + \frac{1}{c}\int \mathrm{d}z' G^{(\mathrm{cp})}(z,z') j'(z') , \tag{4.59}$$

where $A_0(z)$ is the solution of the homogeneous equation (4.49) with $j' = 0$. It constitutes 'incident light' for the j'-induced processes, but it includes the effect of multiple scatterings due to χ_{b} and the linear polarization induced by the QW exciton.

The solution of (4.57) can be obtained as follows. First we apply

$$\left[\frac{\mathrm{d}^2}{\mathrm{d}z^2} + q^2\epsilon_{\mathrm{b}}(z)\right]^{-1}$$

to both sides of the equation, which leads to

$$G^{(\mathrm{cp})}(z,z') + \frac{4\pi}{sc}\left[\frac{\mathrm{d}^2}{\mathrm{d}z^2} + q^2\epsilon_{\mathrm{b}}(z)\right]^{-1} J_{0\nu}(z)\frac{H_{\nu 0}(z')}{[E_{\mathrm{QW}}(k_{\|}) - \hbar\omega - \mathrm{i}\gamma]} = g(z,z') . \tag{4.60}$$

Note the identity

$$\left[\frac{\mathrm{d}^2}{\mathrm{d}z^2} + q^2\epsilon_{\mathrm{b}}(z)\right]^{-1} J_{0\nu}(z) = \frac{-1}{4\pi}\int \mathrm{d}z' g(z,z') J_{0\nu}(z') \quad \left[\equiv \frac{-c}{4\pi} h_{0\nu}(z)\right] . \tag{4.61}$$

The validity of this equation is easily understood by applying

$$\frac{\mathrm{d}^2}{\mathrm{d}z^2} + q^2\epsilon_{\mathrm{b}}(z)$$

to both sides. Multiplying by $J_{\nu 0}(z)$ on both sides of (4.60) and integrating over z, we get

$$H_{\nu 0}(z) + \frac{B_{\nu\nu} H_{\nu 0}(z)}{[E_{\mathrm{QW}}(k_{\|}) - \hbar\omega - \mathrm{i}\gamma]} = h_{\nu 0}(z) , \tag{4.62}$$

where

$$B_{\nu\nu}(\omega) = -\frac{1}{sc^2}\int \mathrm{d}z \int \mathrm{d}z' J_{\nu 0}(z)\, g(z,z') J_{0\nu}(z') , \tag{4.63}$$

$$h_{\nu 0}(z) = \frac{1}{c}\int \mathrm{d}z' J_{\nu 0}(z') g(z',z) . \tag{4.64}$$

Hence, $B_{\nu\nu}(\omega)$ is the radiative correction for the QW exciton and $h_{\nu 0}(z)$ is the vector potential induced by $J_{\nu 0}$. From (4.62) and (4.64), we get

$$H_{\nu 0}(z) = \frac{h_{\nu 0}(z)\left[E_{\mathrm{QW}}(k_{\|}) - \hbar\omega - \mathrm{i}\gamma\right]}{E_{\mathrm{QW}}(k_{\|}) - \hbar\omega - \mathrm{i}\gamma + B_{\nu\nu}} . \tag{4.65}$$

Substituting the above result into (4.60), we obtain

$$G^{(\mathrm{cp})}(z,z') = g(z,z') + \frac{1}{sc}\frac{h_{0\nu}(z)h_{\nu 0}(z')}{E_{\mathrm{QW}}(k_\parallel) - \hbar\omega - \mathrm{i}\gamma + B_{\nu\nu}} \; . \tag{4.66}$$

Hence, the analytic expression for $G^{(\mathrm{cp})}(z,z')$ is obtained in terms of $g(z,z')$ and the induced current density of the QW exciton. The lateral dispersion of the QW exciton is taken into account via $E_{\mathrm{QW}}(k_\parallel)$.

It should be noted that the poles of the second term of (4.66), i.e., the ω s satisfying

$$E_{\mathrm{QW}}(k_\parallel) - \hbar\omega - \mathrm{i}\gamma + B_{\nu\nu}(\omega) = 0 \; , \tag{4.67}$$

give the SS (self-sustaining) modes of the system comprising the coupled QW and DBRs, i.e., cavity polaritons. If we quantize the vacuum photons in a large box, the poles of $g(z,z')$ correspond to the vacuum photons disturbed by the DBR, and the poles of the whole Green function $G^{(\mathrm{cp})}(z,z',\omega)$ to the vacuum photons disturbed by the cavity and the QW exciton.

The sum of the two terms contains poles for the modified vacuum photons and cavity polaritons. The former (latter) has a certain amount of amplitude leaking into (out of) the cavity. In this way, the Green function $G^{(\mathrm{cp})}(z,z')$ describes the propagation of both the modified vacuum photons and the cavity polaritons. By changing ω or $k_\parallel$, we can move smoothly between the vacuum photon and the cavity polariton regimes. There is actually no clear boundary between the two regimes. If we consider a uniform current density with a given $\boldsymbol{k}_\parallel$ at $z = -\infty$ representing the light source, and the cavity occupying the region $0 \le z \le d_{\mathrm{cav}}$, $G^{(\mathrm{cp})}(z = 0^-, z' = -\infty, \boldsymbol{k}_\parallel, \omega)$ gives the reflection amplitude due to the cavity polariton.

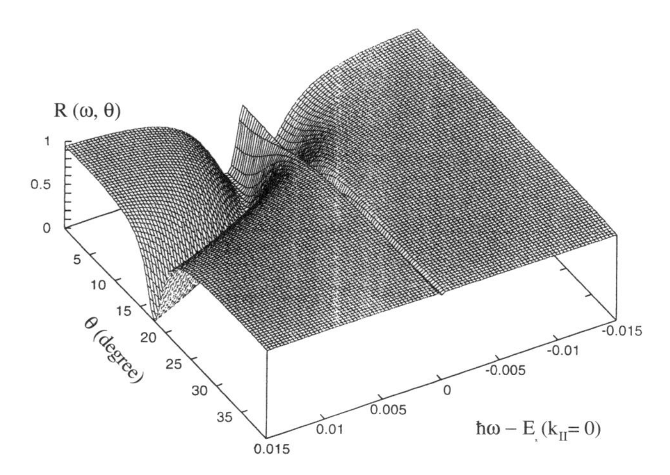

Fig. 4.29. Reflectivity spectrum of a cavity polariton as a function of incident angle, calculated from $G^{(\mathrm{cp})}(z,z',\omega)$. The cavity comprises a pair of DBRs consisting of 14 sets of dielectric bilayers, and the case of zero detuning is shown

Figure 4.29 shows the reflectivity spectrum $R(\omega, \boldsymbol{k}_\parallel)$ calculated in this way. The double dip structure for each $\boldsymbol{k}_\parallel$, representing the upper and lower branches of the cavity polariton, is clearly visible. By choosing various models for the cavity Green function, we can adjust the leakage of the cavity and vacuum modes.

This Green function can be applied to various problems. In particular, it will be interesting in the analysis of the off-branch cavity polaritons reported in [101], in view of the fact that there are rather large signals at frequencies not lying on the cavity polariton branches. The Green function given above will be very useful for discussing the relative intensity and radiative widths of the on- and off-branch signals based on the microscopic model of the cavity. The standard 'quasi-mode coupling Hamiltonian' used in their analysis is a phenomenological approach, and could hardly be justifiable from first principles in the presence of on- and off-branch modes. The present formalism allows us to put the whole analysis on a more rigorous footing.

4.3 Resonant SNOM

Near-field optics is now a very popular subject in a variety of fields including physics, chemistry, biophysics, and engineering. Its aim is not only spectroscopy but also mechanical manipulation of small matter systems, which makes use of the inhomogeneous field produced by light, typically in a near-field region or focused laser beam [104, 105].

The behavior of the EM field in the near-field region of a matter system is rather different from that in the far-field region, which is more familiar to us. For example, it is not the electric dipole component of the EM field that has the largest amplitude. Multipole components can have comparable or even stronger amplitudes in the near field.

Scanning near-field optical microscopy (SNOM) is a new type of microscopy developed in analogy with STM (scanning tunneling microscopy), where the tunneling current between a probe tip and a sample is measured as a function of the tip position. In the case of SNOM, the intensity of light is measured in several different operating modes, specified according to the means of excitation and detection. A typical probe tip is made of a sharpened optical fiber with metal coating, and the very end of the tip has an aperture through which light comes in and/or out. Another type of probe tip is a metal or dielectric sphere, which scatters the EM field produced by the incident light in the vicinity of a sample [106, 107].

The various ways of irradiating a sample and collecting the signal define the different operating modes, such as the irradiation mode (irradiation via probe tip and detection in the far field), collection mode (the inverse process of the irradiation mode), and reflection mode (both irradiation and detection via the probe tip). Among them, we emphasize the special role of the (internal) reflection mode. The specificity of this operating mode is that both the excitation and observation of a sample are made via the near field. In this operating mode, it is possible to excite higher multipole polarization in the sample and to detect the emitted multipole radiation field. If we use plane waves either for excitation or for detection, it is not possible to observe the contribution of the multipole components of induced polarization, either because they are not excited in the sample (collection mode), or because the emitted multipole radiation does not reach the far-field region (irradiation mode). This will be discussed later in this section.

The problems of SNOM are usually considered within the framework of the macroscopic local response. That is, the matter system (sample, substrate and probe tip) is specified by dielectric constant(s) and a given shape, which completely define the Maxwell equations to be solved. Since the light is not resonant with any excitation levels, we may use macroscopic dielectric constants. Theoretical study in such a situation is a challenge, not so much for the physics, but for the numerical computations with a complicated set of boundary conditions. Appropriate numerical methods have been developed in this context [108].

On the other hand, if a resonance is involved, theoretical investigation is more concerned with physical problems such as the origin of the resonance, ways of describing it microscopically, and new features in the observed signal arising from it. Since the resonant frequency sensitively depends on the size and shape of a sample, we cannot simply ascribe a bulk dielectric function to MS or NS systems. Instead, we need a microscopic treatment of the radiation–matter interaction in such systems. Thus the theoretical study of resonant SNOM in nanostructured materials requires us to use the microscopic nonlocal response theory, as in the other examples of this chapter.

The study of resonant SNOM is interesting for several reasons. In the first place, the signal intensity can be stronger than in the non-resonant case. Secondly, since the induced polarization reflects the quantum mechanical details of each excitation level, resonant SNOM measurements at various frequencies are a new kind of spectroscopy in $(\boldsymbol{r}, \omega)$ space, which is a richer source of information than the sample shape.

4.3.1 Configuration Resonance

Some examples of observations making use of resonant conditions reveal

- the tilt angles of individual adsorbed molecules with anisotropic shape [109],
- a very sharp resonance in the position dependence of the surface plasmon signal [110].

In the former example, each molecule is resonantly excited by polarized light from a SNOM tip and the luminescence is detected in the far field. In the latter example, a metal tip and a corrugated metallic surface have a localized plasmon mode with the frequency of the probe light at a particular position of the tip. This has stimulated theoretical work [41] which has demonstrated the configuration resonance effect in terms of a point dipole model. This study has been further developed [111–113] in the form of a model consisting of finite-sized spheres with resonant levels, by applying the original and revised versions of the nonlocal response theory sketched in Sect. 3.4.2.

As repeatedly stressed in previous sections, the signals of resonant optical processes are the emitted EM field from the induced current densities of the whole system, which are determined self-consistently with the vector potential. The larger the amplitudes of the induced current densities, the stronger the signal. When the probe frequency is resonant, the amplitude of the induced current density is generally high.

An important aspect of resonant SNOM is that the response of sample, probe tip and substrate must be solved simultaneously, which means that the resonance condition depends on the position of the probe tip. In particular, the energy eigenvalues of the matter system, as well as their radiative shifts and widths, depend on the position of the probe tip. This leads to the possibility of resonant enhancement of the SNOM signal. If we scan the probe tip in the neighborhood of a sample with a fixed probe light frequency, we may hit a resonance, at a certain position, which will induce a large current density and then a large amplitude of emitted light. If the energy width of the resonance is small, the resonance in the probe position dependence will also be very sharp. Thus we may obtain a very sharply enhanced signal in the spatial mapping of the SNOM signal.

The resonance effect will help to enhance the signal intensity, but the image obtained in the presence of such an effect can be strongly distorted from the geometrical shape of a sample. Thus, it is not appropriate to use the effect for observing the sample shape. Rather, it should be used for observing the details of the electronic structure of a sample whose shape is more or less known.

Let us consider a model system consisting of dielectric spheres on a semi-infinite substrate [112, 113]. We ascribe a single sphere to the probe tip, and an assembly of the same spheres to the sample. Each sphere is assumed to have a resonant level (E_{a} for the sample and E'_{a} for the probe tip), and the substrate has a dielectric constant ϵ_{b}. The transition dipole density corresponding to the excitation in each sphere is expressed as in (4.2). Hence, we have a three-fold degenerate excitation level with x, y, and z polarizations. The current densities are confined in the sphere and there is no overlap between those belonging to different spheres. The induced current densities on the spheres interact with one another via the EM field. The contribution of the L component of the EM field produces the instantaneous dipole–dipole interaction among all the induced current densities. On the other hand, the T component contributes to the radiative interaction among the current densities, which causes radiative width as well as radiative shift in each excitation energy. In the presence of a semi-infinite substrate with a macroscopic di-

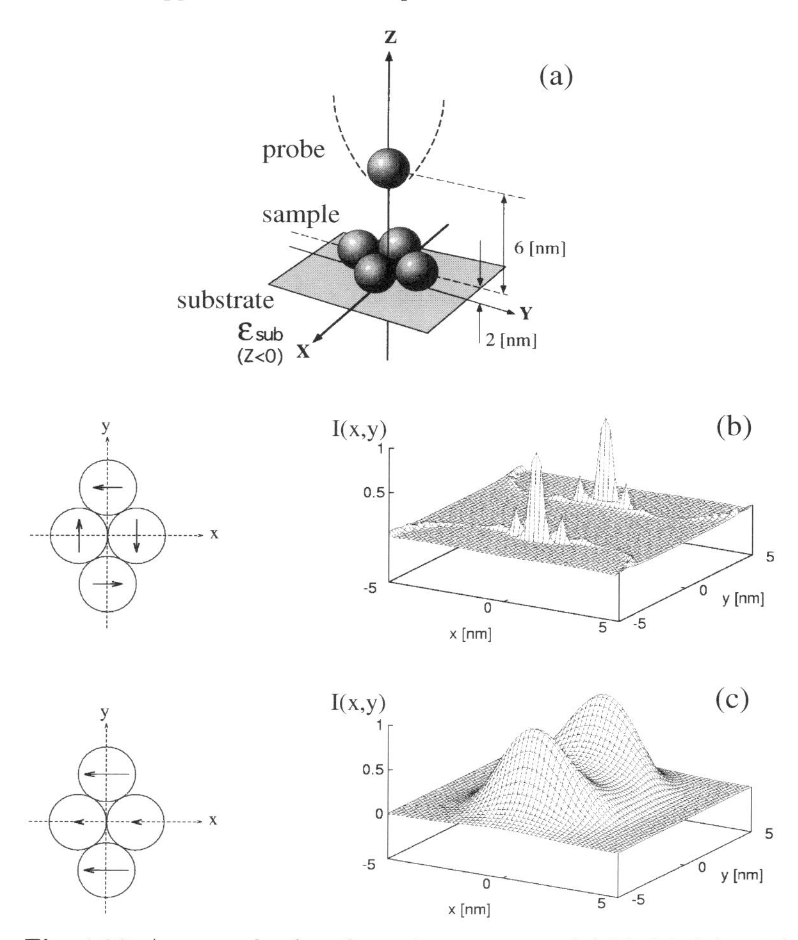

Fig. 4.30. An example of configuration resonance. (**a**) Model, (**b**) and (**c**) intensity of induced polarization at the probe sphere for the two modes specified by the polarization patterns on the left-hand sides defined at $x = y = 0$

electric constant $\epsilon_{\rm b}$, it is convenient to use the renormalized Green function $\mathbf{G}_{\rm r}^{(\rm M)}$ of the EM field. The eigenmodes of the whole system depend on the position (x, y) of the probe tip.

Figure 4.30b and c shows two examples at $(x = 0, y = 0)$. On the left-hand side, polarization patterns are shown, and on the right-hand side, the corresponding SNOM image, i.e., the polarization intensity on the probe tip as a function of its position. All the sharp structures in Fig. 4.30b are due to configuration resonance, reflecting the very small spectral width of the mode.

4.3.2 Breakdown of the Dipole Selection Rule in Reflection Mode

It is commonly accepted that important optical transitions among various quantum mechanical states are electric dipole (ED) transitions, and that other multipole transitions only occur as minor or negligible contributions. Although this is quite a widespread belief in the traditional framework of the optical response, it should be remembered that it is based on the LWA (long wavelength approximation), i.e., it is assumed that the spatial variation of the EM field inducing transitions in a sample is negligible compared with that of the relevant matter wave functions.

It is possible to think of a situation which does not satisfy the LWA condition. When we consider resonant transitions in a mesoscopic structure whose size is comparable to or exceeds the resonant wavelength, there is no justification for using LWA. Furthermore, the wavelength to be compared with the sample size is not the one in vacuum, but the one determined self-consistently. As shown in several cases in this chapter, such self-consistency generally leads to the dominance of short wavelength components. Hence, the condition for LWA to be valid becomes more severe. A typical example of this is the 1s exciton in a slab of CuCl, discussed in Sect. 4.1.2 (Fig. 4.10). The internal field calculated self-consistently has a large spatial variation for a slab thickness of about 280 Å and an incident energy of 3.202 eV, whose wavelength in vacuum is about 4 000 Å. Such a remarkable growth of short wavelength components in a resonant field generally occurs within the nonlocal framework, and it can be regarded as a new guiding principle for obtaining large nonlinear signals, where the EM field can have a large spatial variation. This will be discussed in more detail in Chap. 5.

There is another type of situation where the EM field is not constant in the region covering a sample, even if the sample size is much smaller than the resonant wavelength. This is when the field just outside the aperture of a SNOM tip (which we assume to have a size smaller than the sample) has a large spatial variation over the length scale of the tip size. If a sample is irradiated by such light, the field acts locally on the sample. This means that any excited states having appreciable transition current density in that local area can be effectively excited. Thus any multipole excited states may be excited by a SNOM tip if it is positioned appropriately.

On the other hand, the excited multipole emits a corresponding multipole radiation field, which has characteristic radial and angular dependence according to each mode. In the far field, the ED component is dominant. In the near field, however, higher multipole fields can have appreciable amplitudes, so that they have a chance of being observed effectively by a SNOM tip.

Combining the above arguments, we can expect to excite and observe the eigenmodes of matter with higher multipole components by using a SNOM tip for both irradiation and signal detection. This is usually called the reflection mode of SNOM operation. The name arises from the fact that we make use of a change in the internal reflectivity of an optical fiber in the presence and absence of a sample at the other end of the fiber. By changing the position of the SNOM tip with respect to the sample, the internal reflectivity changes.

We show the possibility of detecting the ED and higher multipole modes of a sample with comparable amplitudes in terms of a simple model calculation below [114].

As a sample, we consider a 1D chain of 10 QDs (quantum dots) along the z-axis. Each one has radius R and the nearest neighbor distance is b_0. Each QD has an ED allowed resonant level E_0, and the excitation energy can transfer from QD to QD via dipole–dipole interaction. The total length of the chain ($10b_0$) is assumed to be much smaller than the resonant wavelength hc/E_0. To simulate a SNOM tip, we consider an additional sphere of QD, which has a slightly different resonant level. Corresponding to the positions of the sample spheres, $(0, 0, nb_0)$, where $n = 0, 1, 2, \ldots, 9$, the probe-tip sphere is assumed to be swept parallel to the z-axis at a given distance h.

The incidence of light is simulated by giving an initial EM field on this (probe-tip) sphere alone, and the signal is defined as the field produced by the self-consistent polarization on this sphere alone. (As a slightly more realistic image of this probe-tip, one might imagine a SNOM tip with an additional QD, which is connected to an optical fiber.) By calculating the signal intensity as functions of the tip position and light frequency, we can compare the relative strength of various multipole transitions which have different resonance energies.

Figure 4.31 shows the polarization patterns of the eigenstates of the chain for the two highest levels, with polarization perpendicular to the chain. They are distinguished by the number of nodes and the mirror symmetry with respect to the center of the chain. The state with an odd number of nodes has an ED component, and the other does not. The transition dipole moment for the latter is zero. If one fixes the tip position at the point where the polarization of the mode takes the maximum value, the resultant spectrum of the signal intensity of the reflection mode SNOM defined above looks like those in Fig. 4.32. On the other hand, the signal intensity as a function of probe tip position is given in Fig. 4.33 for each excited state, for a fixed frequency. The sharp double line structure is due to the configuration resonance effect,

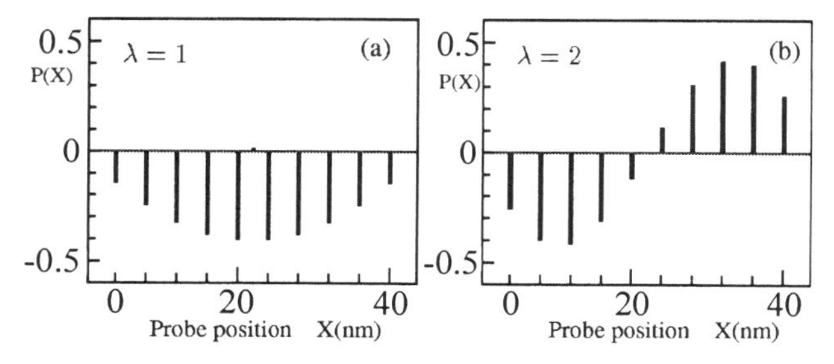

Fig. 4.31a,b. Polarization patterns of the two eigenmodes of the 1D chain

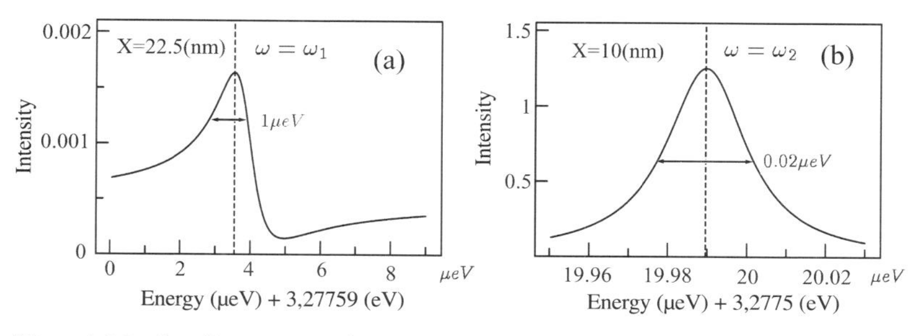

Fig. 4.32a,b. Spectrum of each mode for a fixed position X of the probe tip. Reprinted from [114], with permission from Elsevier

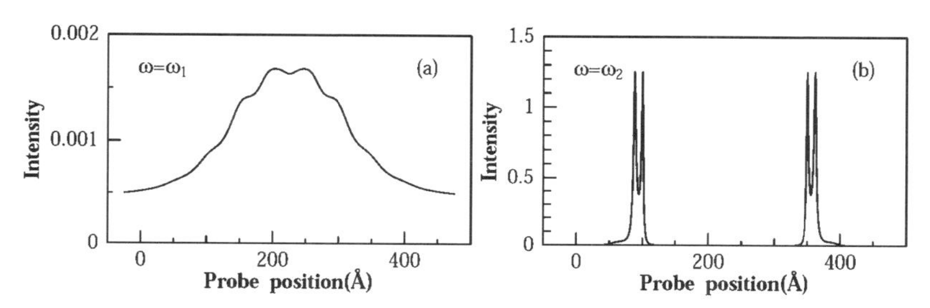

Fig. 4.33a,b. SNOM pattern (signal intensity) at a fixed frequency, i.e., the vertical lines in Fig. 4.32. Reprinted from [114], with permission from Elsevier

i.e., as the tip position is changed, the eigenfrequency of the coupled sample plus probe-tip system is changed, and this can produce a sharp resonance effect with the fixed probe frequency at nearby points. Since a resonance produces a large polarization on the probe sphere, we can expect a large signal. This situation is shown in Fig. 4.34 for the two modes ω_1 and ω_2. The sharp resonance occurs for the ω_2 mode, while the resonance is dull for the ω_1 mode.

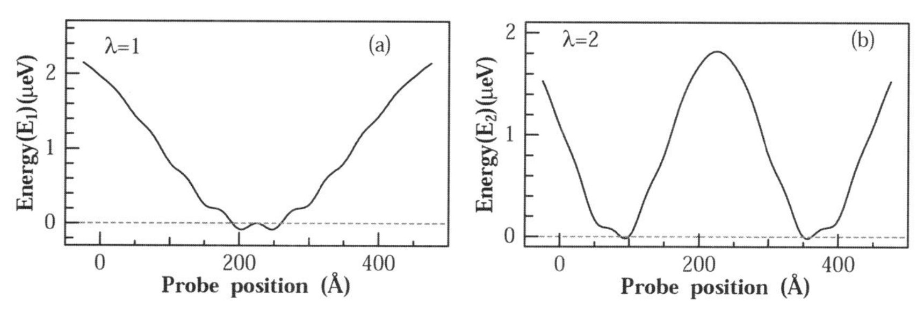

Fig. 4.34a,b. Probe position dependence of the mode energies for $\lambda = 1, 2$. Reprinted from [114], with permission from Elsevier

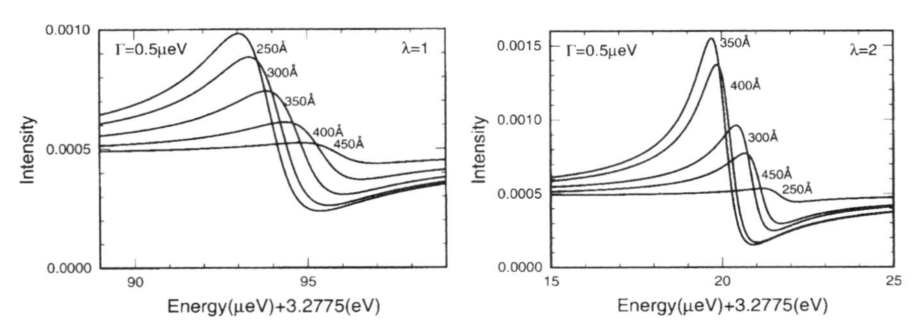

Fig. 4.35. Integrated signal intensities in the presence of non-radiative damping $\Gamma = 0.5$ μeV for modes $\lambda = 1, 2$ at various probe positions

The width of the peak in Fig. 4.32 is due to the radiative one, since no non-radiative damping is assumed in the calculation. Reflecting the oscillator strength of each mode, the width is quite different from mode to mode. Although the presence of the probe tip can in principle break the dipole-forbidden character of the states with even-number nodes, the result shows that this effect is not appreciable. If one compares the integrated intensity of the peaks in Fig. 4.32, the ω_1 and ω_2 modes make nearly the same contribution, which confirms the aim of this model calculation. Hence, the resonant SNOM intensities of the ED allowed and forbidden modes can be comparable, if the tip position is set at the point of maximum polarization of the mode. For more realistic purposes, one would like to include the non-radiative width of the excited matter levels. Figure 4.35 shows the spectra of the two modes for various tip positions. Here also, one can see the comparable signal intensities of the two modes.

4.4 New Method of Photonic Band Calculation

The general dispersion equation (3.53) given in Sect. 3.6 has various features extending the known formulations of polaritons, X-ray dynamical scattering, and photonic crystals. The generalized feature depends on the problem to be studied. As a polariton dispersion, it contains the effect of reciprocal lattice scattering, i.e., the effect of short wavelength components of the induced polarization. As a dynamical scattering problem with X rays, it contains the effect of resonant levels. In this section, we will be concerned with the problem of photonic bands.

The most popular way to calculate photonic band structure is the plane wave (PW) expansion. Here one solves the set of Fourier-transformed Maxwell equations. Eliminating the electric field $\boldsymbol{E}_{\boldsymbol{k}+\boldsymbol{g}}$ (magnetic field $\boldsymbol{H}_{\boldsymbol{k}+\boldsymbol{g}}$) from the equations, where $\boldsymbol{g}$ represents the reciprocal lattice vectors, one obtains a set of linear equations for $\{\boldsymbol{H}_{\boldsymbol{k}+\boldsymbol{g}}\}$ ($\{\boldsymbol{E}_{\boldsymbol{k}+\boldsymbol{g}}\}$). The coefficients contain various Fourier components of the inverse dielectric constant (dielectric constant) with periodic structure. The solutions $\{\omega_j(\boldsymbol{k})\}$ of these equations determine the photonic band structure. The method for solving this set of equations is called the H-method [115] (E-method [116]), depending on the remaining variables.

Although both sets of equations are equivalent for an infinite set of $\{\boldsymbol{g}\}$, they are not equivalent in reality, where one cannot use infinitely many $\boldsymbol{g}$ s for numerical calculation. Their difference is expected to tend to diminish as one uses more and more $\boldsymbol{g}$ s. However, convergence is very slow and often accompanied by oscillatory behavior. It is often the case that several thousand PWs, a typical number for ordinary computers, are not enough to guess the convergence points [117].

It is therefore desirable to develop a different method with better convergence behavior and thus with higher precision for ordinary computers. For this purpose, (3.53) in Sect. 3.6 is a good candidate, because of the structure of the equation. Although it also makes use of the PW expansion, the matrix to be diagonalized is the inverse of the one used in the H-method, as is obvious from the fact that the eigenvalues for a fixed $\boldsymbol{k}$ are not q^2, but $1/q^2$. The truncated part of the matrix due to the use of a finite number of $\boldsymbol{g}$ s can be made much smaller than those of the E- or H-method. This is because each matrix element in (3.53) contains $|\boldsymbol{k}+\boldsymbol{g}||\boldsymbol{k}+\boldsymbol{g}'|$ in the denominator, while the same factor appears in the numerator in the E- or H-method. Since these factors become larger and larger in the truncated part, it is not clear whether the truncated matrix is a good approximation to the original one for determining the eigenvalues in the lower frequency range. In this respect, the inverse matrix is easier to work with [45, 46].

However, there is an additional step involved in solving (3.53). Indeed, we have to determine $\bar{\chi}_{\boldsymbol{k}}(\boldsymbol{g},\boldsymbol{g}')$ from the corresponding elements of the ordinary susceptibility $\chi_{\boldsymbol{k}}(\boldsymbol{g},\boldsymbol{g}')$. The susceptibilities χ and $\bar{\chi}$ are defined with respect to the full Maxwell electric field and its T component, respectively, and are related to each other as in (3.58). The spatial structure of a photonic crystal is defined through the $\boldsymbol{r}$ dependence of the local $\chi(\boldsymbol{r})$, whose $(\boldsymbol{g},\boldsymbol{g}')$ dependence is a function of $\boldsymbol{g}-\boldsymbol{g}'$ alone. This dependence is another advantage for the inverse matrix diagonalization, because a function of $\boldsymbol{g}-\boldsymbol{g}'$ requires much less memory size than a function of $(\boldsymbol{g},\boldsymbol{g}')$.

In this way, we can formulate a new high-precision band calculation scheme on the basis of (3.53) [46]. For a given model of the photonic crystal, we prepare all the necessary $\chi(\boldsymbol{g}-\boldsymbol{g}')$ s. Let us write the tensor components of (3.58) as

$$\bar{C}_{ij} = C_{ij} - C_{i3}(1+C_{33})^{-1}C_{3j} \,, \tag{4.68}$$

where

$$C_{ij}(\boldsymbol{g},\boldsymbol{g}') = \hat{\mathrm{e}}_i(\boldsymbol{k}+\boldsymbol{g})\cdot 4\pi\chi_{\boldsymbol{k}}(\boldsymbol{g},\boldsymbol{g}')\cdot\hat{\mathrm{e}}_j(\boldsymbol{k}+\boldsymbol{g}') \,. \tag{4.69}$$

The definition of $\bar{C}_{ij}$ is made similarly by replacing χ by $\bar{\chi}$. The inverse matrix $(1+C_{33})^{-1}$ is evaluated via power series expansion.

Because of the good behavior of the truncated matrix to be diagonalized, we do not need a very large matrix for $\bar{C}(\boldsymbol{g},\boldsymbol{g}')$. Let its size be $2N \times 2N$, where $N \leq 100$ will be enough to calculate the lower frequency bands. The factor of two arises from the number of transverse directions for each wave vector $\boldsymbol{k}+\boldsymbol{g}$. To determine this small matrix for each combination of (i,j), we use the $2N \times 2M$ matrix of $C(\boldsymbol{g},\boldsymbol{g}')$, where M should be taken as large as possible. This M is the maximal number of PWs to be considered in the band calculation, and it is possible to take a value of 10 000 with ordinary computers. It should be stressed once again that the memory size for this $2N \times 2M$ matrix is not $2N \times 2M$, but only M because of its $(\boldsymbol{g}-\boldsymbol{g}')$ dependence.

Once the small matrices $\bar{C}_{ij}(\boldsymbol{g},\boldsymbol{g}')$ are determined for a given $\boldsymbol{k}$, we can calculate the eigenvalues $1/q^2$, and this gives the photonic band dispersion $\{\omega_j(\boldsymbol{k})\}$.

Figure 4.36 shows the comparison of the M dependence of the lowest photonic gap between the H-method and the new method mentioned above for the model of an intersecting square rod structure. This shows the stable convergence behavior of the present method.

The main problem with the new method is not the memory size, but the time limit for evaluating the inverse matrix via a power series expansion, i.e., how many expanded terms should be used for a given precision in the result. There is a more important problem, concerning the question as to whether this series expansion is really convergent. For this problem, we can introduce an additional parameter, and adjust it so that all the eigenvalues of the expanded matrix have amplitudes less than 1. For most of the models we tested, this condition turned out to be satisfied, which guaranteed convergence. But there is no proof that convergence is always guaranteed. For such a model, we

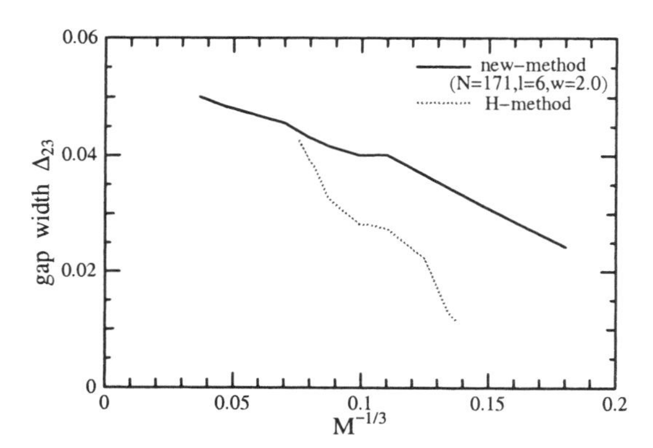

Fig. 4.36. The width of the lowest photonic gap as a function of the (effective) number of plane waves. *Continuous* and *dotted curves* are the result of the new method and the H-method, respectively. Reprinted from [46], with permission from Elsevier

cannot apply the new method. However, we have not yet encountered such a malicious case.

4.5 Resonant Photonic Crystals

In Sect. 3.6 we presented a general dispersion equation for the EM field in periodic material systems. It can be applied to general cases of photonic bands even in the presence of resonant levels. This has contributed to the development of a new method of photonic band calculation based on diagonalization of the inverse matrix, as explained in the previous section. Effectively it enables us to include more plane wave basis sets than the usual plane wave expansion method in numerical calculation, so that the resultant band structure becomes more reliable.

Photonic crystals have been intensively studied, mainly from an engineering point of view [118]. This is probably because the fundamental physics of the problem is well understood, while its applicational possibilities remain open and promising. However, the targets of such studies are almost exclusively restricted to those consisting of non-resonant materials, which requires the macroscopic local description of the EM field. Photonic crystals with electronic resonances are poorly studied systems. Since they may require microscopic nonlocal description of the EM field, the fundamental physics of these systems remains to be explored, as does their potential for applications.

In this section, an example of such a resonant photonic crystal is given to show that there are still controversial points concerning the understanding of very fundamental aspects, such as the number of photonic bands. This may indicate the need to study resonant photonic crystals. The problem concerns the resonant Bragg reflector described in Sect. 4.1.4, which consists of a periodic array of N identical layers, with an electronic resonance with spacing d satisfying the Bragg condition for normally incident light at the resonant wavelength.

The dispersion relation for the photonic bands of this system can be obtained from the general expression (3.53) of Sect. 3.6 [119]. However, we can take a simpler approach in view of the simple mathematical structure due to the 1D character of the system with only one resonant level in each unit cell (for a given $\boldsymbol{k}_\parallel$) [85]. In the following, we do not write $\boldsymbol{k}_\parallel$ explicitly, because it is a good quantum number for both the matter and the EM field.

For $N = \infty$, we also have periodicity in the normal direction, so that the excited matter states are characterized by the wave number κ in the normal direction. In the present model, there is only one band of excited states of the matter, i.e., for each wave number κ there is only one eigenstate. Then the general condition for the eigenmodes $\det|\mathbf{S}| = 0$ discussed in Sect. 3.2 is factorized into the 1×1 matrix equations for the individual κ:

$$0 = E(\kappa) - \hbar\omega + \mathcal{A}_{\kappa 0, 0\kappa}(\omega) \,, \tag{4.70}$$

where $\mathcal{A}_{\kappa 0,0\kappa}(\omega)$ is the radiative correction calculated for the resonant level with wave number κ.

Evaluating the radiative correction (4.39) in Sect. 4.1.4 in the present case for a given $\boldsymbol{k}$ ($\boldsymbol{k}_{\|} = 0, \kappa$), we obtain

$$\mathcal{A}_{\kappa 0,0\kappa}(\omega) = -\frac{4\pi}{\Omega c^2} \sum_{g} \frac{|S(\kappa+g)|^2}{(\kappa+g)^2 - (q+\mathrm{i}0^+)^2} , \tag{4.71}$$

where g is the reciprocal lattice vector of the 1D lattice, $S(\kappa+g)$ the Fourier component of the current density, and Ω the volume of the unit cell.

This expression together with (4.70) gives us the desired equation for the resonant photonic bands. We put $g = g_\ell = 2\ell\pi/d$, ($\ell = 0, \pm 1, \pm 2, \ldots$), and $E(k) = \hbar\omega_0$, neglecting the κ dependence of the resonant energy $E(\kappa)$. The Bragg condition $d = \lambda_0/2$ is rewritten as $d = \pi c/\omega_0$ or $g_1/2 = \omega_0/c$ ($= q_0$, the boundary of the first Brillouin zone). The dispersion equation of the present model is

$$\hbar c(q_0 - q) = \frac{|S(0)|^2}{\kappa^2 - q^2} + \sum_{\ell=1}^{\infty} |S(2\ell q_0)|^2 \left[\frac{1}{(\kappa + 2\ell q_0)^2 - q^2} + \frac{1}{(\kappa - 2\ell q_0)^2 - q^2} \right] . \tag{4.72}$$

The solution $q = q(\kappa)$ of this equation gives the dispersion relation of the EM wave in this resonant photonic crystal. This form of the equation is convenient for graphical solution. We draw the right- and left-hand sides of (4.72) as functions of q ($= \omega/c$) in the neighborhood of the lowest photonic gap. The crossing points of the curves of the right-hand side with the straight line of the left-hand side give the solutions with real k and ω. Obviously, there are three real roots in this frequency region. The solutions for all the possible κ give us the dispersion curves, as shown in Fig. 4.37. The gap mode has very weak dispersion, lying between ω_0 and $\omega_0 + |\Delta_0|$, where Δ_0 is the radiative shift of a single layer for $\boldsymbol{k}_{\|} = 0$.

The above argument is general enough to convince us of the existence of the gap mode. However, there exists a more frequently used approach to the dispersion relation based on the so-called transfer matrix method, according to which no report has been made of the existence of the gap mode for the same resonant Bragg condition. The reason for the disagreement was unknown for a while. However, it turns out to be essential to keep the very small term of the radiative shift Δ_0 in the denominator of the reflection and transmission amplitudes leading to the transfer matrix elements, in order to obtain a similar gap mode branch numerically.

On the other hand, while studying the analytical equivalence of the two schemes, an incomplete feature of the transfer matrix method came out. In fact, equivalence is guaranteed by the use of reflection and transmission amplitudes which differ in certain respects from those in the literature [86]. Although the difference is negligible in the limit of vanishing thickness of

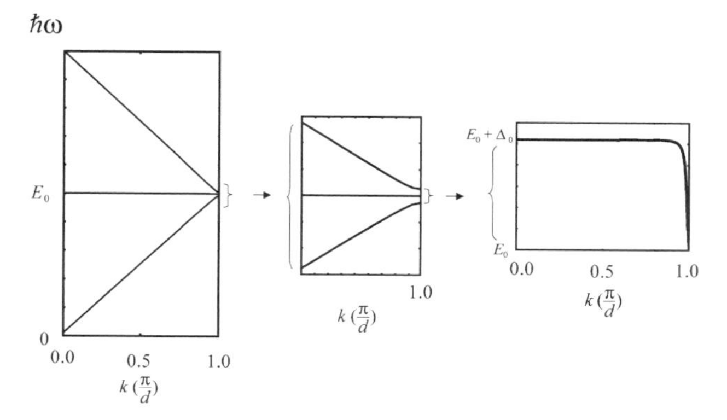

Fig. 4.37. Photonic band dispersion of a 1D resonant Bragg reflector near the lowest gap for $k_{\|} = 0$. Δ_0 is the radiative shift of a single layer. The gap mode is shown on an expanded scale

each layer, it is generally necessary to use different forms of the reflection and transmission amplitudes in calculating transfer matrix elements.

Let us outline how we might show the equivalence of the two schemes. In the transfer matrix method, we relate the right- and left-going field amplitudes $\{E_j^{(+)}, E_j^{(-)}\}$ between the neighboring interlayer positions (j and $j+1$) via the transfer matrix elements, which are independently calculated for a single-layer problem. On the other hand, the field amplitudes in the nonlocal scheme are the sum of contributions from all the layers. The weight of the emitted light from each layer X_j is determined from the fundamental equations for this scheme, viz., $\mathbf{S}\tilde{\boldsymbol{X}} = \tilde{\boldsymbol{F}}^{(0)}$. If the two schemes are equivalent, the response field should be the same, so that the equations for $\{E_j^{(+)}, E_j^{(-)}\}$ rewritten in terms of $\{\tilde{X}_j\}$ should coincide with $\mathbf{S}\tilde{\boldsymbol{X}} = \tilde{\boldsymbol{F}}^{(0)}$.

The transfer matrix elements are determined from the reflection (r_1) and transmission (t_1) amplitudes of a single layer. The solution of this single-layer problem in the nonlocal framework gives

$$r_1 = \frac{2\pi \mathrm{i}}{qc^2} \frac{\mathcal{I}_{01}^*(q)\mathcal{I}_{01}(-q)}{E_0 - \hbar\omega + \tilde{A}_{10,01}}, \tag{4.73}$$

$$t_1 = 1 + \frac{2\pi \mathrm{i}}{qc^2} \frac{\mathcal{I}_{01}^*(q)\mathcal{I}_{01}(+q)}{E_0 - \hbar\omega + \tilde{A}_{10,01}}, \tag{4.74}$$

where the Fourier coefficient of the current density in the layer is

$$\mathcal{I}_{\mu\nu}(q) = \int \mathrm{d}z \mathrm{e}^{-\mathrm{i}qz} I_{\mu\nu}(z) \tag{4.75}$$

for the lateral component of the current density accompanying the transition from $|\nu\rangle$ to $|\mu\rangle$. Using the Fourier transform of the 1D Green function, we can evaluate $\mathcal{A}$ as

$$\mathcal{A}_{10,01} = \frac{2\pi}{qc^2} \int \mathrm{d}k \left(\frac{1}{k+q+\mathrm{i}0^+} - \frac{1}{k-q-\mathrm{i}0^+} \right) |\mathcal{I}_{01}(k)|^2 \,. \tag{4.76}$$

Its imaginary part $-\Gamma_0$ is given by

$$\Gamma_0 = \frac{\pi}{qc^2} \left[|\mathcal{I}_{01}(q)|^2 + |\mathcal{I}_{01}(-q)|^2 \right] \,. \tag{4.77}$$

Ivchenko's result is

$$r_1' = \mathrm{i} \frac{\Gamma_0}{E_0 - \hbar\omega + \tilde{A}_{10,01}} \tag{4.78}$$

with $t_1' = 1 + r_1'$. Comparing (4.73) and (4.78), we find that $2\mathcal{I}_{01}^*(q)\mathcal{I}_{01}(-q)$ is replaced by $|\mathcal{I}_{01}(q)|^2 + |\mathcal{I}_{01}(-q)|^2$ in r_1'. This approximation is valid for $q \to 0$, in other words, for vanishing thickness of the resonant layer. The reflection and transmission amplitudes in (4.73) and (4.74) do not satisfy the relation $t_1 = 1 + r_1$. However, it is amusing to see that they satisfy the expected normalization $|r_1|^2 + |t_1|^2 = 1$ exactly, although this is not obvious at first sight.

If we use the new amplitudes r_1 and t_1 given above by the transfer matrix method, we can show the equivalence of the equations for the field components of the two schemes analytically [120].

A peculiar feature of the gap mode becomes evident in its optical response spectrum and the corresponding internal field patterns. The spectral range of this mode lies in the middle of the photonic gap, i.e., the total reflection range. This means that a narrow window opens at the center of the total reflection region for the frequency range of the gap mode. In this range, the internal field should have a pattern of propagating modes, instead of the decaying pattern of evanescent waves in the total reflection region. This can be seen quite clearly in the following calculations for the finite array of a resonant Bragg reflector [85]. Because of the finite thickness, the propagating

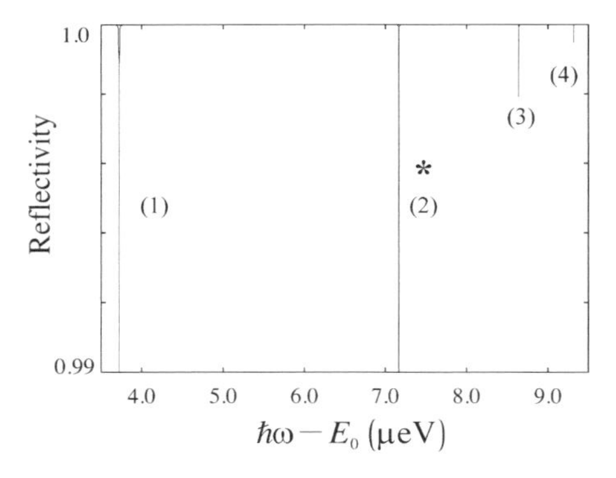

Fig. 4.38. Dips in the total reflection range of 1D a resonant Bragg reflector with $N = 101$ layers

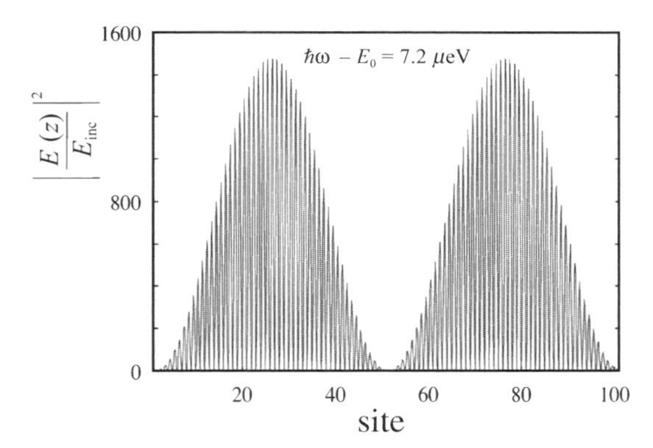

Fig. 4.39. Internal field pattern at the frequency marked by an asterisk in Fig. 4.38

waves of the gap mode are quantized, forming standing wave patterns inside the system.

Figure 4.38 shows the very narrow dips in the middle of the total reflection region corresponding to the quantized levels of the gap mode. The internal field patterns at these dip frequencies are shown in Fig. 4.39. They vividly demonstrate the standing wave nature of the quantized gap modes. In these calculations, non-radiative decay was neglected for simplicity, so that the Q-factor of the quantized gap modes can be extremely large. Its inclusion will significantly reduce the Q-factors. Its effect on the response in frequency and time domains is an open question at the time of writing.

5. Application: Nonlinear Response

The theoretical description of nonlinear response has been well documented since the early book by Bloembergen [121]. However, the case of resonant phenomena, which often needs a self-consistent determination of the EM field and polarization, does not seem to have been satisfactorily studied. In particular, the nonlocality of nonlinear response has been very poorly investigated. For this reason, we put our main stress on nonlocal aspects of the problem in the present chapter on nonlinear response. This will provide us with a new mechanism for nonlinear processes, which can be used as a new guiding principle for inventing nonlinear devices.

Bloembergen et al. have considered the nonlocality of the radiation–matter interaction in a nonlinear context in their theory of second harmonic generation (SHG) from a medium with inversion symmetry [122]. They consider nonlocality by using the spatial derivatives of the EM field as well as the EM field itself to describe the components of the induced current density. Instead of considering the position-dependent current density on the nanometer scale, they use macroscopic variables for the oscillating electric dipole and quadrupole moments and the rotation of magnetization, which interact with the electric field, electric field gradient, and magnetic field, respectively. They thereby analyzed SHG processes in media with inversion symmetry, demonstrating the importance of the surface-type quadrupole term. This approach is a limiting case of the nonlocal theory. Hence, if the induced current density is a weakly varying function of space, we can expand it in a Taylor series at each (macroscopic) poin t and keep the first few terms which correspond to the three components mentioned above. A complete generalization should include the MS or NS spatial variation of the induced current density in more detail, and this leads to the microscopic nonlocal scheme described in the present book.

In the standard description of nonlinear processes, we need nonlinear susceptibilities, in addition to the linear susceptibility. Susceptibilities of various orders can be calculated quantum mechanically for a given microscopic model of the matter system. In each step of the perturbation calculation, we must evaluate the matrix elements of the radiation–matter interaction. There are three such terms, as discussed in Chap. 2, and for resonant processes the main contribution comes from the term

$$H_{\rm int} = -\sum_{\ell} \frac{e}{mc} \boldsymbol{p}_\ell \cdot \boldsymbol{A}(\boldsymbol{r}_\ell) \ . \tag{5.1}$$

It is common practice to evaluate this using LWA, i.e.,

$$\langle \mu | H_{\rm int} | \nu \rangle \approx -\langle \mu | \sum_{\ell} \frac{e}{mc} \boldsymbol{p}_\ell | \nu \rangle \cdot \boldsymbol{A}(\boldsymbol{r}) \ , \tag{5.2}$$

with the assumption that $\boldsymbol{A}(\boldsymbol{r})$ has a slow spatial variation. One considers only the electric dipole transitions as the main contributions to the susceptibilities. This approximation is known to be valid as long as the spatial extension of the relevant quantum mechanical states $\{|\mu\rangle, \ |\nu\rangle\}$ is smaller than the wavelength of the (resonant) light.

A question arises at this stage: what happens if the LWA condition is no longer valid? This question has not been seriously considered in the study of optical response in general, except for certain special cases such as the ABC problem discussed in Sect. 3.8.

In the case of linear response, we have shown that the EM field determined self-consistently with the induced polarization can contain components with strong spatial variation under resonant conditions, even if its wavelength in vacuum is much longer than the size of the matter. If we choose an optimal condition with respect to sample size and light frequency, the dominant component can be the one with strong spatial variation, rather than the one with long wavelength. Since the EM field obtained in this manner can act as a source field to induce nonlinear polarization, there arises a new situation where the nonlinear polarization is induced by a spatially varying field. In this case, we have to evaluate the full matrix element of $\langle \mu | H_{\rm int} | \nu \rangle$, i.e., $\tilde{F}_{\mu\nu}/c$ of (2.75), instead of (5.2) based on LWA. In particular, some or all of the elementary steps of quantum mechanical transition producing a nonlinear polarization may become different from the electric dipole-type transition.

In nonlinear optics, this is a totally different situation from the one commonly considered. It results from the self-consistency of the EM field and polarization and the use of nonlocal susceptibilities. As for the self-consistency, it may also be required for resonant processes in the macroscopic (local) response, where, however, the spatial variation of the EM field near a resonance does not contain information concerning the characteristic spatial variation of the quantum mechanical wave function of the resonant level. This is because each resonance is characterized by an eigenfrequency and oscillator strength in the macroscopic theory. As the frequency approaches a resonance, the wave number of the EM field increases, but the spatial variation generally bears no resemblance to that of the corresponding wave function.

On the other hand, the microscopic polarization directly reflects the form of the wave function of the resonant state, and it bears a nonlocal relationship with the source EM field within the spatial extension of the resonant state. The polarization thus reflects the characteristic feature of each resonant level, in both strength and spatial form, and it is transferred to the EM field through the requirement of self-consistency.

In considering nonlinear processes according to the microscopic nonlocal scheme, the main emphasis will be put on the resonant enhancement of nonlinear signals as functions of sample size and light frequencies. The enhancement is caused, not only by the energy denominator of the susceptibility, but also by the amplitude of the (self-consistently determined) internal electric field with characteristic spatial patterns. The latter aspect bears a certain resemblance to Fabry–Pérot interference in the macroscopic response theory based on LWA, but it can be more dramatic in amplitude and spatial variation when we go beyond LWA. A remarkable point is that the spatially varying internal field with large amplitude leads to the possibility for 'non-electric-dipole' type transitions to make the main contribution to nonlinear polarization.

It should be stressed that, while the enhancement in macroscopic theory is almost always argued with respect to susceptibility, we talk about signal intensity in microscopic nonlocal theory, because it is the result of the two enhancement mechanisms for susceptibility and internal field.

A more specific argument can be made as follows. The signal intensity is determined by nonlinear polarization, which is a sum of the integrals over relevant coordinates of the nonlinear susceptibility multiplied by certain numbers of internal field components. As shown in Sect. 2.6, the nonlinear polarization (of the third order for example) contains the products of three $\tilde{F}_{\mu\nu}$ factors divided by various frequency factors. The factors $\tilde{F}_{\mu\nu}$ are ω dependent, having resonant enhancement at a certain matter excitation energy (plus its radiative correction), as we frequently discussed in the linear response cases. In view of the fact that matter excitation energy and its radiative correction depend on the size, shape and internal structure of a sample, the nonlinear signal can be enhanced via two different factors, i.e., the energy denominators and the factors $\tilde{F}_{\mu\nu}$ of nonlinear polarization [55, 70, 71, 123]. Because the resonant enhancement can be obtained by adjusting light frequencies and sample size, we may call this effect Nonlocality-Induced DOuble Resonance in Energy and Size (NIDORES).

The NIDORES effect should be clearly distinguished from the size enhancement of nonlinear susceptibility described in Sect. 3.9. The latter is related to the linear size enhancement of dipole matrix elements appearing in the nonlinear susceptibility within LWA, and the enhancement tends to be saturated when the system size becomes comparable to the resonant wavelength. On the other hand, the former becomes conspicuous beyond LWA. Although both mechanisms are size-sensitive, one is size-linear and the other is size-resonant. They are both MS or NS effects, but the essential features of nonlocal response are reflected only in the size-resonant NIDORES effect.

We regard it as very important to establish the NIDORES effect both theoretically and experimentally, because it is a clear indication of the importance of the microscopic nonlocal viewpoint in the description of resonant optical processes for MS or NS matter systems. This will certainly contribute to establishing the hierarchical structure of the theoretical frameworks for the radiation–matter interaction, i.e., QED on the most fundamental level, the microscopic nonlocal scheme on the next, and the macroscopic local scheme on the last. (As we move up these levels, we need more approximation at each step.) This structure may be obvious for a small number of experts, but there has been little firm experimental evidence for the need for the nonlocal theory. The experimental results to be mentioned in Sect. 5.2 will contribute to this purpose.

The NIDORES effect in resonant nonlinear processes is expected to work in most confined systems. For the explicit explanation of the effect in this chapter, we will treat the case of weakly confined excitons in a slab, with normally incident pump and probe lights, where the only exciton and bound two-exciton (biexciton) states relevant to the linear and nonlinear processes are those with vanishing lateral wave vector, i.e., $\boldsymbol{k}_\parallel = 0$. For the process of degenerate four-wave mixing (DFWM), we take the limit of normal incidence after choosing the right term in $\chi^{(3)}$. Although our demonstration is limited to this simple situation in this chapter, the effect itself can be expected quite generally in MS or NS systems, where radiative shift and level separation can be comparable.

5.1 Pump–Probe Spectroscopy

This type of spectroscopy may be considered as one in which we probe (by probe light) matter systems disturbed by the pump light. As mentioned in the introduction to this chapter, we will be concerned with resonant processes for excitons and biexcitons confined in a slab. Hence, both the probe and pump light frequencies (denoted $\omega_{\rm pr}$ and $\omega_{\rm pu}$, respectively) are chosen in resonance with some of the transition energies of the matter system. We denote the size-quantized energies of the exciton and biexciton measured from the ground state energy by $(E_{{\rm x}n})$ and $(E_{{\rm b}j})$, respectively, and consider two cases:

$$\begin{aligned} &(\mathrm{A})\ \ \hbar\omega_{\rm pu} \approx E_{{\rm x}n}\ , \quad \hbar\omega_{\rm pr} \approx E_{{\rm b}j} - E_{{\rm x}n}\ , \\ &(\mathrm{B})\ \ \hbar\omega_{\rm pr} \approx E_{{\rm x}n}\ , \quad \hbar\omega_{\rm pu} \approx E_{{\rm b}j} - E_{{\rm x}n}\ . \end{aligned}$$

In case (A), the ground state of the matter is mixed with the exciton state by the pump light, which leads to a certain real population of the excited state. Such mixing of the exciton and ground states will allow further transitions to biexciton states induced by the probe light. Since this additional absorption channel occurs in the neighborhood of the linear exciton absorption, the effect of the pump light is to modify the linear absorption in the energy range of the exciton excitation. One would expect a pump-induced absorption.

On the other hand, case (B) corresponds to the optical Stark effect, where two excited states, mixed by the pump light, will cause a change in the linear absorption in the energy range of the exciton excitation. In this case, one would expect two-photon absorption to biexciton levels, the optical Stark effect, where the linear absorption peaks are shifted in a characteristic way, and also a pump-induced broadening of the linear absorption peaks.

In both cases, the resonances in the third-order susceptibility are given as the differences of the eigenenergies of the biexciton, exciton and ground states. Considering that the resonant energies in the bulk for the exciton (E_x) and exciton–biexciton transition ($E_b - E_x$) differ by the binding energy Δ_{bx} of the biexciton, we may assume that the resonant poles contributing to cancellation in the nonlinear susceptibility $\chi^{(3)}$ are separated by about Δ_{bx}. If the non-radiative width of the resonance is much smaller than Δ_{bx}, which is the case in the following examples, we may consider that the cancellation problem discussed in Sect. 3.8 is not serious, as long as we confine ourselves to the resonant processes. With this situation in mind, we pick up only resonant terms in $\chi^{(3)}$ for calculation of the nonlinear response spectrum.

The third-order term in the induced current density in the presence of the pump and probe light can be written (2.121). Since the nonlinear signal amplitude is directly proportional to this current density, we see two ways to enhance the signal intensity:

(a) make use of the resonant condition on the denominators for various frequency combinations,
(b) enhance the factors $\{\tilde{F}(\omega)\}$ in the numerators.

The equations determining the factors $\{\tilde{F}(\omega)\}$ are (2.88) corrected with cubic terms. However, as far as the perturbation scheme is appropriate, the essential part of the solution is determined by the linear part, i.e., $\tilde{\boldsymbol{F}} = \mathbf{S}^{-1}\tilde{\boldsymbol{F}}^{(0)}$, which is proportional to $[\det(\mathbf{S})]^{-1}$. Hence, the factors $\{\tilde{F}(\omega)\}$ have resonant enhancement at the poles of $[\det(\mathbf{S})]^{-1}$, which are the matter excitation energies with radiative corrections. Since the radiative shift in MS or NS systems can be comparable to the intervals of size quantization energies, mechanisms (a) and (b) for the enhancement of nonlinear signals work differently with respect to the light frequencies and sample size. The optimal condition between the two mechanisms gives the NIDORES effect for the pump–probe process, to be described in detail in the next section [55, 70, 71].

5.1.1 Pumping of Exciton Absorption

By choosing resonance condition (A) mentioned above, we pick up the most resonant terms of $\chi^{(3)}$ with respect to the energy denominators, which contribute to the third-order polarization with the probe light frequency ω_{pr}. Because of the relation (2.120) for the frequencies of the source and induced fields, the pump frequency ω_{pu} occurs twice and the probe frequency ω_{pr} once ($\omega_{pr} = \omega_{pr} + \omega_{pu} - \omega_{pu}$) in the pump–probe process among the three frequency-dependent factors $\tilde{F}$. With this condition in mind, we examine the terms in (2.121) carefully, and find two types of triply resonant case, where the three factors in the denominator of $\chi^{(3)}$ take the form:

- $(\omega_{\rm pr} - E_{jn})(\omega_{\rm pr} + \omega_{\rm pu} - E_{{\rm b}j})(\omega_{\rm pu} - E_{{\rm x}n})$,
- $\pm(\omega_{\rm pr} - E_{jn})(\omega_{\rm pu} - \omega_{\rm pu} - E_{{\rm x}n} + E_{{\rm x}n'})(\omega_{\rm pu} - E_{{\rm x}n})$,

where $E_{jn} = E_{{\rm b}j} - E_{{\rm x}n}$. The first case arises from the second term of (2.121) with the choice

$$\begin{aligned} &\Omega_3 = \omega_{\rm pu} , \quad \Omega_2 = \omega_{\rm pr} + \omega_{\rm pu} , \quad \Omega_1 = \omega_{\rm pr} , \\ &E_\sigma = E_\mu = E_{{\rm x}n} , \quad E_\nu = E_{{\rm b}j} . \end{aligned} \tag{5.3}$$

The second case comes from the fourth or fifth term of (2.121), with the choice

$$\begin{aligned} &\Omega_3 = \omega_{\rm pu} , \quad \Omega_2 = \omega_{\rm pu} - \omega_{\rm pu} , \quad \Omega_1 = \omega_{\rm pr} , \\ &E_\sigma = E_\mu = E_{{\rm x}n} , \quad E_\nu = E_{{\rm b}j} , \end{aligned} \tag{5.4}$$

in the case of the fourth term, or

$$\begin{aligned} &\Omega_3 = -\omega_{\rm pu} , \quad \Omega_2 = \omega_{\rm pu} - \omega_{\rm pu} , \quad \Omega_1 = \omega_{\rm pr} , \\ &E_\sigma = E_\mu = E_{{\rm x}n} , \quad E_\nu = E_{{\rm b}j} , \end{aligned} \tag{5.5}$$

in the case of the fifth term.

Although each factor goes to zero at an exact resonance in these expressions, they are broadened in realistic models due to the presence of damping mechanisms. The effect of non-radiative damping can be introduced phenomenologically, by assuming damping terms in the equations of motion for the components of the matter density matrix [124]. As an example of such an approach, we may introduce two (non-radiative) damping constants γ and Γ for population decay and phase decay between the ground and one-exciton states, respectively. The phase decay constant between the ground and biexciton states is assumed to be 2Γ. Then the contribution of the above two terms to the third-order nonlinear polarization gives

$$\frac{|\tilde{F}_{n0}(\omega_{\rm pu})|^2 \tilde{F}_{jn}(\omega_{\rm pr})}{(\omega_{\rm pr} + {\rm i}\Gamma - E_{jn})(\omega_{\rm pr} + \omega_{\rm pu} + 2{\rm i}\Gamma - E_{{\rm b}j})(\omega_{\rm pu} + {\rm i}\Gamma - E_{{\rm x}n})} , \tag{5.6}$$

and

$$-\frac{2\Gamma|\tilde{F}_{n0}(\omega_{\rm pu})|^2 \tilde{F}_{jn}(\omega_{\rm pr})}{(\omega_{\rm pr} + {\rm i}\Gamma - E_{jn})\gamma[\Gamma^2 + (E_{{\rm x}n} - \omega_{\rm pu})^2]} . \tag{5.7}$$

These are the most resonant terms with respect to the energy denominators under condition (A). Another type of resonant enhancement arises from $\tilde{F}$s in the numerators. The linear susceptibility χ, which is the main factor determining the ω dependence of $\tilde{F}(\omega)$, has resonance for $\omega_{\rm pu}$ but not for $\omega_{\rm pr}$ under condition (A). We therefore expect only the factor $|\tilde{F}_{n0}(\omega_{\rm pu})|^2$ to be a source for this type of enhancement.

In the slab geometry, the factor $\tilde{F}_n(\omega)$ is the expansion coefficient of the internal electric field with respect to the wave functions $\{\sin(k_n z)\}$, where

$k_n = n\pi/(N+1)d$, $n = 1, 2, \ldots, N$, represents the wave number for the center-of-mass motion of the confined exciton. (The slab is assumed to consist of N atomic layers with lattice constant d, i.e., the slab thickness is Nd.)

In order to understand the results concerning nonlinear response in this section, it is quite helpful to refer to the linear response of the slab system mentioned in Sect. 4.1.2. Figure 4.9 shows the N and ω dependence of the factor $|\tilde{X}_n(\omega)|^2$ for several n values. This reveals its sensitive dependence on ω and the slab thickness Nd. The ridges in the figure showing resonant enhancements correspond to the poles of $\det(\mathbf{S})_{\mathrm{x}}^{-1}$, and they cross the curves representing the size-quantized exciton energies $E_{\mathrm{x}n}$ as functions of N. Since $(\omega_2 - E_{\mathrm{x}n}/\hbar)$ is one of the resonant factors in the triply resonant terms of $\chi^{(3)}$ for the pump–probe process mentioned above, the crossing points are candidates for double enhancement with respect to frequency and size.

For the detailed calculation of nonlinear effects, we took a simplified model of a slab [71], namely, a bundle of linear chains in the normal direction to the slab, with the chain length being the thickness of the slab. Excitons are confined on each N site chain, and the interaction among them is considered only between the nearest sites on the same chain. This allows the formation of bound two-exciton states, i.e., biexcitons. Thus the model Hamiltonian includes three parameters, i.e., the site energy of the exciton E_0, the nearest-neighbor transfer energy $-b$, and the nearest-neighbor interaction $-\bar{\delta}$. The eigenvalues and wave functions of the biexcitons are calculated numerically. The numerical values of the parameters are chosen so as to reproduce the linear spectrum and the biexciton binding energy of the Z_3 exciton in CuCl. Although we neglect interchain transfer and scattering states with $\boldsymbol{k}_{\parallel} = 0$ consisting of two excitons with $\pm\boldsymbol{k}_{\parallel,1}$, this does not affect the qualitative significance of the conclusion, i.e., the existence of nonlocality-induced resonance enhancement of the nonlinear signal in the mesoscopic range of sample sizes.

Figures 5.1 and 5.2 show the pump-induced change in the transmittance of the probe light through the slab as functions of the probe light frequency ω_{pr} and either the pump light frequency ω_{pu} or the slab thickness Nd. In Fig. 5.1, N is fixed at 50 (layers), and in Fig. 5.2, $\hbar\omega_{\mathrm{pu}}$ is fixed at $E_{\mathrm{x}1}$. If $\hbar\omega_{\mathrm{pu}}$ is fixed at the resonant peak associated with $E_{\mathrm{x}2}$, i.e., the excitation energy with radiative correction, we obtain the curves in Fig. 5.3. In these figures, one can observe the remarkable enhancement of the pump-induced transitions between $E_{\mathrm{x}j}$ and $E_{\mathrm{b}\mu}$.

The nonlinear signal shows resonance with respect to both the probe light frequency ω_{pr} and either the pump light frequency ω_{pu} or the slab thickness Nd. This fact reflects the two resonance mechanisms underlying the enhancement, i.e., the energy denominator of $\chi^{(3)}$ or the factors $\{\tilde{F}\}$. The enhancement factor becomes very large for the smaller value of the non-radiative damping $\varGamma = 0.06$ meV. This value of $\varGamma$ is realistic, since it was derived from analysis of the ultrafine interference pattern measured for a

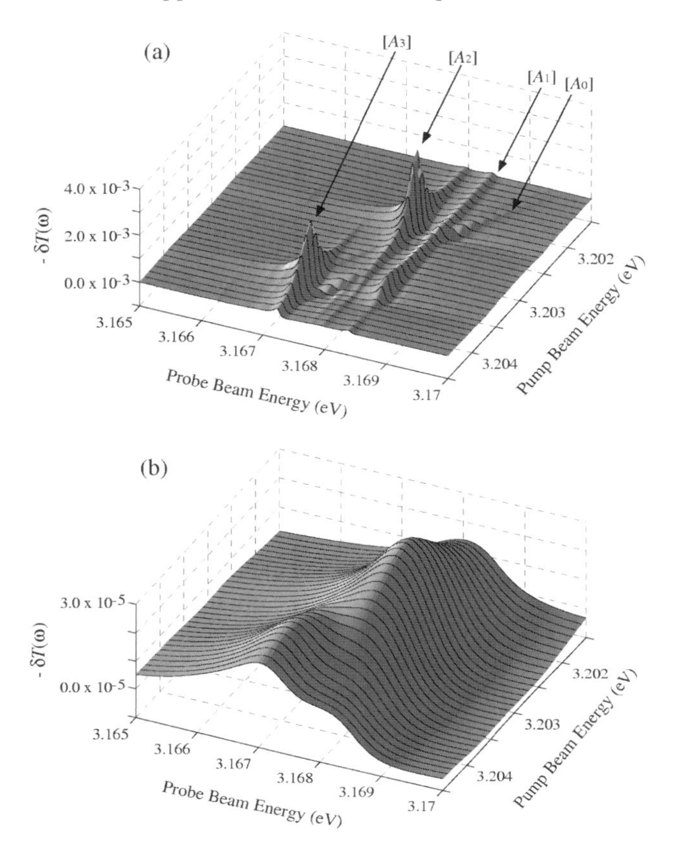

Fig. 5.1. Pump energy dependence of the change in transmittance spectrum $-\delta T(\omega_{\rm pr})$ for size $N = 50$. The values of (γ, Γ) in [meV] are (**a**) (0.02, 0.06), (**b**) (0.2, 0.6) [71]

very thin slab in the resonant region of the Z_3 exciton in CuCl [52, 125]. The enhancement is sensitive to the values of damping constants, as can be seen by comparing the curves in Figs. 5.1–5.3.

The enhanced signals denoted by $[A_j]$ in Fig. 5.1 are the transitions from $E_{{\rm x}j}$ to $E_{{\rm b}j}$, respectively. Each of them has a peak in its $\omega_{\rm pu}$ dependence, due to the resonant enhancement of $\tilde{F}_j(\omega_{\rm pu})$. For these $\omega_{\rm pu}$ values, the induced current density contains a large amplitude due to excitation of the $E_{{\rm x}j}$ exciton state, and this contributes to the second optical transition to the $E_{{\rm b}j}$ biexciton state by the probe light. Since the probe light frequency is off-resonant with respect to the exciton transition, its spatial structure is very flat and the main final state from the $E_{{\rm x}j}$ exciton state is therefore the j th biexciton state (i.e., the transition in LWA). The difference in their peak positions and sizes is due to the different conditions for the resonant enhancement of the

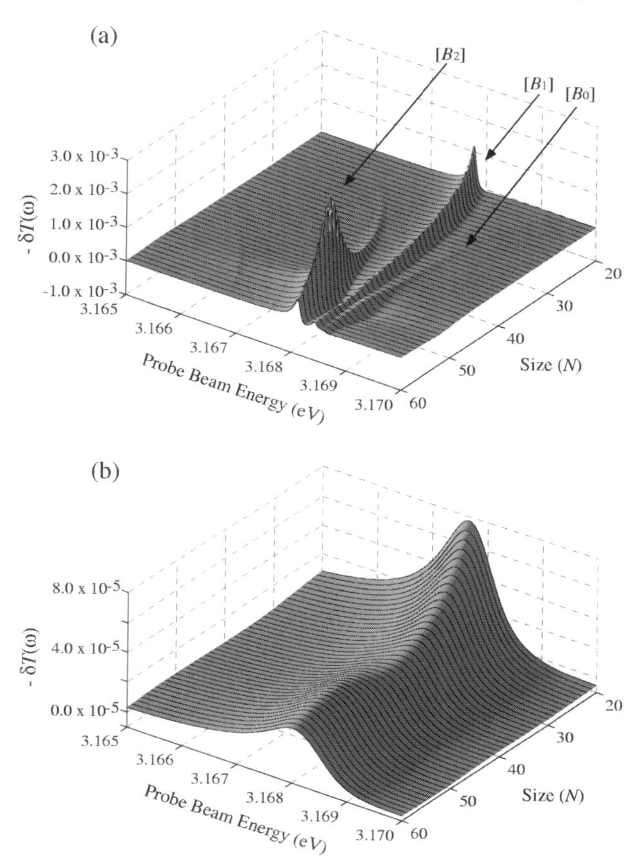

Fig. 5.2. Size dependence of the change in transmittance spectrum $-\delta T(\omega_{\rm pr})$ for a fixed $\omega_{\rm pu} = E_{\rm x1}$. The values of (γ, Γ) in [meV] are (**a**) (0.02, 0.06), (**b**) (0.2, 0.6) [71]

factor $\tilde{F}_j(\omega_{\rm pu})$ for $N = 50$. A qualitatively different resonant behavior is seen in Fig. 5.1 , denoted by $[A_0]$. This is the two-photon transition satisfying $\hbar(\omega_{\rm pr} + \omega_{\rm pu}) = E_{\rm b1}$. Although this resonance is not the one mentioned above explicitly, such a term exists among $\chi^{(3)}$.

In Fig. 5.2, there are resonant structures denoted by $[B_j,\ j = 0, 1, 2]$. Their peak energies satisfy the relationships $\hbar\omega_{\rm pr} = E_{\rm b2} - E_{\rm x1}$, $E_{\rm b1} - E_{\rm x1}$, $E_{\rm b2} - E_{\rm x2}$ for $j = 0, 1, 2$, respectively. Considering that $\hbar\omega_{\rm pu} = E_{\rm x1}$ in Fig. 5.2, we can interpret them as follows. B_0 is the two-photon absorption to the level $E_{\rm b2}$, B_1 is a superposition of the two-photon absorption to $E_{\rm b1}$ and the induced transition, like A_1 via the pump-induced population of the exciton level $E_{\rm x1}$, and finally, B_2 has the characteristic behavior of the double-resonance enhancement of the transition A_2, whose size resonance occurs at $N = 52$. The size dependence of B_2 well reflects that of $|\tilde{F}_2|^2$ in Fig. 4.9.

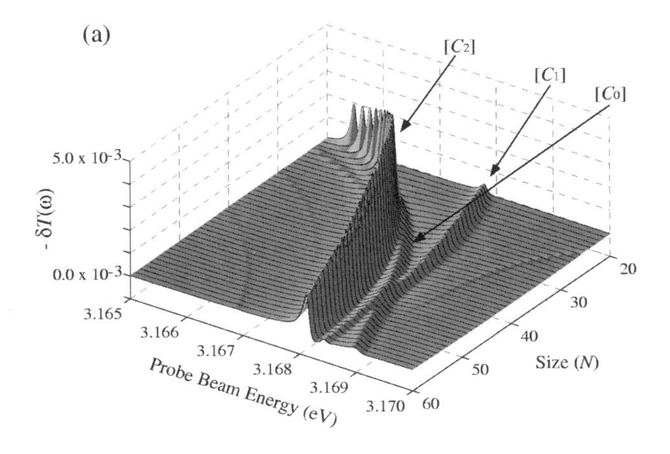

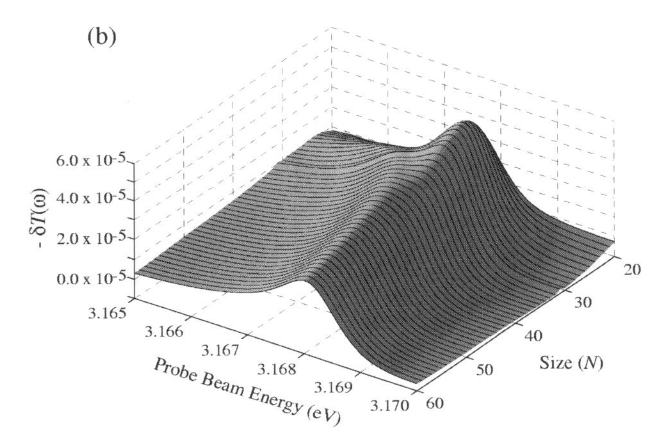

Fig. 5.3. Size dependence of the change in transmittance spectrum $-\delta T(\omega_{\rm pr})$ for a fixed $\omega_{\rm pu}$ at the second resonance energy of the linear response spectrum. The values of (γ, Γ) in [meV] are (**a**) (0.02, 0.06), (**b**) (0.2, 0.6) [71]

5.1.2 Pumping of Exciton–Biexciton Transition

For the condition (B) given in the first part of Sect. 5.1, the pump frequency as a linear process is off-resonant from the exciton energy by the biexciton binding energy. Therefore, a resonant enhancement of the factor $|F_{j\rm x}(\omega_{\rm pu})|^2$ cannot be expected. As mentioned in the introduction to this chapter, the interesting physical feature in this case is not the same as in the previous case. What we have here is a distortion of the exciton energy scheme due to the pumping (optical Stark effect), with consequences for the probe signals.

If the binding energy $\Delta_{\rm bx}$ of two excitons in forming a biexciton is much larger than the exciton and biexciton line widths, and if we are interested in the lower members of size-quantized exciton and biexciton levels, where size-quantization energies are smaller than $\Delta_{\rm bx}$, the factors $(\Omega_{{\rm x}n} - \omega_{\rm pu} - {\rm i}\gamma)$ and $(\Omega_{{\rm b}j} - 2\omega_{\rm pu} - {\rm i}\gamma)$ in the denominator of $\chi^{(3)}$ are regarded as non-resonant. The third-order induced current density $\tilde{\boldsymbol{j}}^{(3)}$ with frequency $\omega_{\rm pu}$ arises from the combinations of $\omega_{\rm pr}$ and $\omega_{\rm pu}$ as one of the following cases:

- $(\omega_1, \omega_2, \omega_3) = (\omega_{\rm pr},\ -\omega_{\rm pr},\ \omega_{\rm pu})$,
- $(\omega_1, \omega_2, \omega_3) = (\omega_{\rm pu},\ -\omega_{\rm pu},\ \omega_{\rm pu})$.

However, both cases can be discarded because the probe light has weak intensity in the first, and the energy denominator contains the above-mentioned non-resonant factors in the second. This means that all third-order terms in the frequency $\omega_{\rm pu}$ can be neglected, so that we can determine the EM field of the frequency $\omega_{\rm pu}$ as a linear problem.

On the other hand, $\tilde{\boldsymbol{j}}^{(3)}$ with the probe light frequency $\omega = \omega_{\rm pr}$ contains a triply resonant term

$$\frac{1}{(\hbar c)^3} \sum_{n,n',j} \frac{\langle 0|\boldsymbol{I}(\boldsymbol{r})|{\rm x}n'\rangle \tilde{F}_{{\rm x}n',{\rm b}j}(-\omega_{\rm pu}) \tilde{F}_{{\rm b}j,{\rm x}n}(\omega_{\rm pu}) \tilde{F}_{{\rm x}n,0}(\omega_{\rm pr})}{(\Omega_{{\rm x}n} - \omega_{\rm pr} - {\rm i}\gamma)(\Omega_{{\rm b}j} - \omega_{\rm pr} - \omega_{\rm pu} - {\rm i}\gamma)(\Omega_{{\rm x}n'} - \omega_{\rm pr} - {\rm i}\gamma)} . \tag{5.8}$$

This is obtained from the first term of (2.121) by taking

$$\Omega_3 = \omega_{\rm pr} , \quad \Omega_2 = \omega_{\rm pr} + \omega_{\rm pu} , \quad \Omega_1 = \omega_{\rm pr} + \omega_{\rm pu} - \omega_{\rm pu} ,$$
$$\omega_{\sigma 0} = \Omega_{{\rm x}n} , \quad \omega_{\nu 0} = \Omega_{{\rm b}j} , \quad \omega_{\mu 0} = \Omega_{{\rm x}n'} .$$

We thus solve the problem in the following way. First, we determine the field for $\omega_{\rm pu}$ by solving the linear problem. Then, we obtain the factors $\tilde{F}_{{\rm x}n',{\rm b}j}(-\omega_{\rm pu})$ and $\tilde{F}_{{\rm b}j,{\rm x}n}(\omega_{\rm pu})$ in the triply resonant term of $\chi^{(3)}$ mentioned above, viz., (5.8). At this stage, the factors $\tilde{F}_{{\rm x}n',{\rm b}j}(-\omega_{\rm pu})$ and $\tilde{F}_{{\rm b}j,{\rm x}n}(\omega_{\rm pu})$ are not unknown variables. This term acts like a correction to the linear current density due to the pump light. Adding this approximate term to the linear induced current density for frequency $\omega_{\rm pr}$,

$$\tilde{\boldsymbol{j}}^{(1)}(\boldsymbol{r}, \omega_{\rm pr}) = \sum_n \frac{\langle 0|\boldsymbol{I}(\boldsymbol{r})|{\rm x}n\rangle \tilde{F}_{{\rm x}n,0}(\omega_{\rm pr})}{\Omega_{{\rm x}n} - \omega_{\rm pr} - {\rm i}\gamma} , \tag{5.9}$$

we find the response field for frequency $\omega_{\rm pr}$ in the presence of the pump light. The modified problem has linear form in the sense that one only needs to solve the coupled linear equations for a single frequency $\omega_{\rm pr}$.

The effects of all non-resonant components in the third-order induced current density $\tilde{\boldsymbol{j}}^{(3)}(\boldsymbol{r}, \omega)$ are neglected, and those of the linear current density are treated as a background with an effective dielectric constant $\epsilon_{\rm b}$, as discussed in Sect. 3.4. We now show several numerical results from [126], in which the Maxwell equations with the effective linear susceptibility, including the third-order effect in the above-mentioned way, are solved as the second-order differential equations with appropriate boundary conditions for the Maxwell field. Although the radiative correction is not explicitly evaluated, its effect is included during the process of solution. This is equivalent to solving the fundamental equation of the nonlocal formalism, $\mathbf{S}\tilde{\boldsymbol{F}} = \tilde{\boldsymbol{F}}^{(0)}$, for resonant polarization components alone, with the radiative correction term evaluated in terms of the radiation Green function including the background dielectric constant, as discussed in Sect. 3.4.

We consider a model consisting of a thin layer of semiconductor (CuCl) of thickness d, where the CM motions of excitons and biexcitons are confined in a rigid wall potential. In particular, their CM wave number in the surface normal (z) direction is quantized as $K = n\pi/d$, $n = 1, 2, \ldots$, and the corresponding CM wave function has the form $\phi_{\ell,K} = \sqrt{2/d}\sin(KZ)$ with $\ell =$Ex and Em for the exciton and biexciton, respectively. Assuming normal incidence for both pump and probe beams, we take the eigenenergies of the exciton and biexciton system as $\hbar\omega_\ell(K) = \hbar\omega_\ell(0) + \hbar^2K^2/2M_\ell$, where M_ℓ is the translational mass of the exciton ($\ell =$ Ex) or biexciton ($\ell =$ Em). The matrix elements of the induced current density are

$$\langle\phi_{\mathrm{Ex},K}|\hat{I}(Z)|0\rangle = I_0\sqrt{\frac{2}{d}}\sin(KZ)\ , \tag{5.10}$$

$$\langle\phi_{\mathrm{Em},K'}|\hat{I}(Z)|\phi_{\mathrm{Ex},K}\rangle = J_0\sqrt{\frac{2}{d}}\sin(KZ)\sin(K'Z)\ . \tag{5.11}$$

The matrix elements I_0 and J_0, reflecting the details of the wave functions of the electron–hole relative motions, are treated as parameters in this calculation.

Figure 5.4 shows the dispersion curves of bulk excitons and polaritons (UP and LP) together with the size-quantized exciton levels (at wave numbers $\{k_j\}$) for a slab with $d = 200$ Å. These levels appear as resonant structure in response spectra such as reflection (R), transmission (T) and absorption ($A = 1 - R - T$). In the following, we show the results of the nonlocal calculation in terms of the normalized absorption $A_\mathrm{n} = A/(1 - R)$. Apart from the lower members of the quantized levels, the level positions correspond well to the peaks of $A_\mathrm{n}(\omega)$, as shown in Fig. 5.4.

In Fig. 5.5, we show the change in $A_\mathrm{n}(\omega)$ due to the pump light at the frequencies denoted $\{a_2, a_3, a_4; x_3, x_4, x_5\}$. The positions a_j and x_j correspond to the non-resonant and resonant frequencies of the probe light, respectively. The vertical dashed lines in Fig. 5.5 show the size-quantized levels of the biexciton which can be reached by the sum of the pump and probe light frequencies, i.e., $E_{\mathrm{b}j} = \hbar(\omega_\mathrm{pr} + \omega_\mathrm{pu})$, for each fixed probe light frequency. These are the two-photon absorption processes via exciton states. The effect of two-photon absorption appears differently for ω_pr in resonance or out of resonance with $E_{\mathrm{x}n}$, i.e., in comparison with the linear process, the absorption tail (non-resonant part) increases and the absorption peak (resonant part)

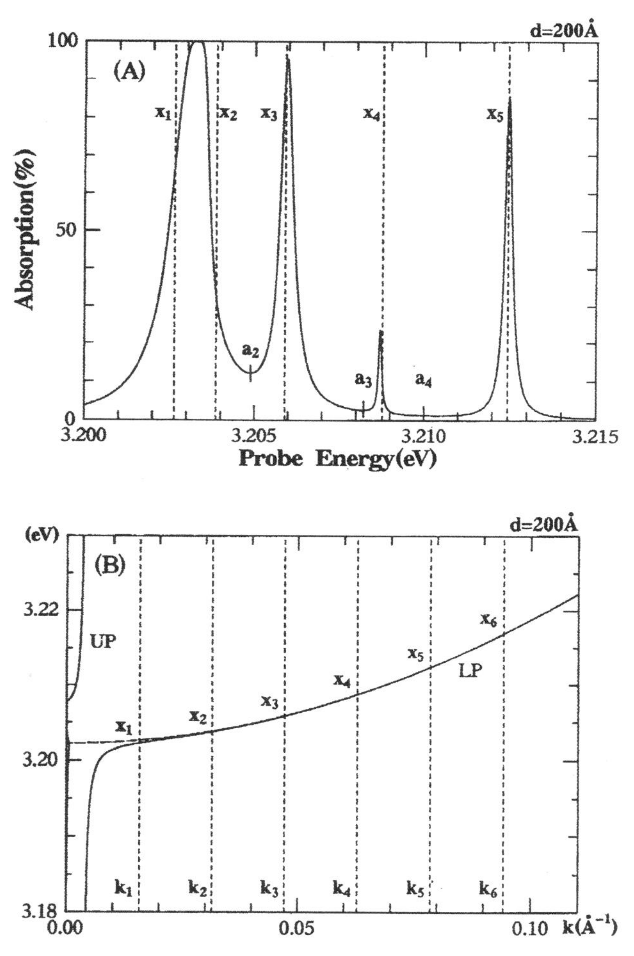

Fig. 5.4. (**A**) Normalized absorption spectrum of a CuCl slab with $d = 20$ nm and (**B**) polariton dispersion curves. *Vertical lines* in (**B**) indicate the size-quantized wave numbers, $\{x_i\}$ are the energy positions of size-quantized excitons, and $\{a_i\}$ the absorption minima [126]

decreases. The decrease is due to the induced broadening of the absorption peak in the presence of the pump light.

In view of the fact that the pump field has weak spatial structure, the pump-induced transition between the exciton and biexciton occurs almost vertically in the wave number space, i.e., the intermediate exciton level of the transition to $E_{\mathrm{b}m}$ is mostly $E_{\mathrm{x}m}$. The intensity ratio of these transitions reflects the amplitude ratio of the exciton states contributing to the polarization induced by the probe light. This can be checked independently by calculating the amplitude ratio as a function of ω_{pr}.

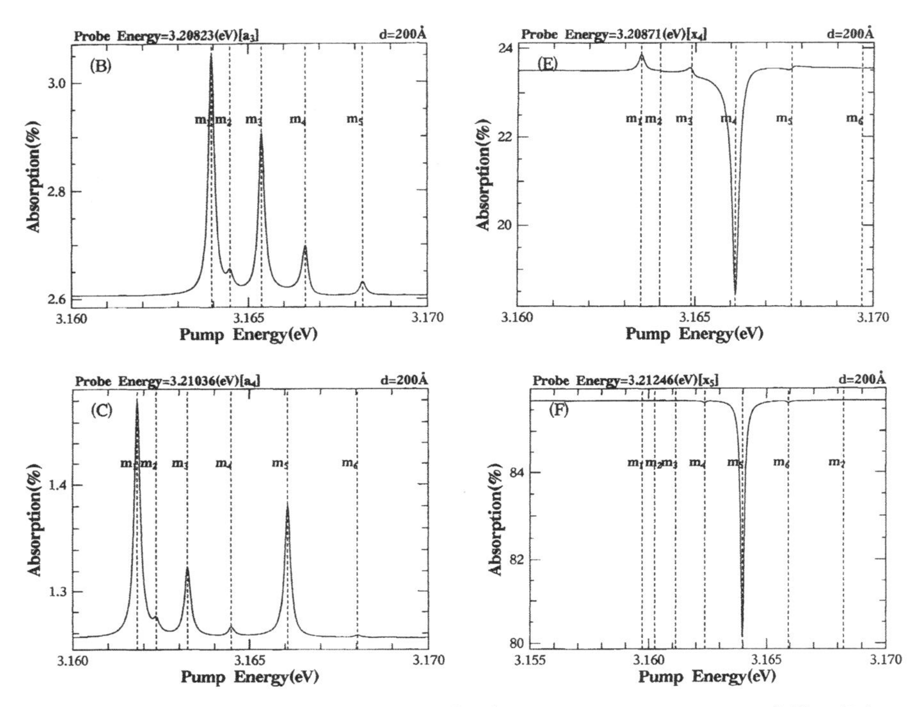

Fig. 5.5. Change in absorption at probe frequencies a_3, a_4, x_4, x_5 of Fig. 5.4 as a function of ω_{pu}. *Vertical lines* denoted m_j indicate the pump energy satisfying $\hbar\omega_{\text{pu}} = E_{\text{bx}}(k_j) - \hbar\omega_{\text{pr}}$. The value of $\hbar\omega_{\text{pr}}$ is given in each frame [126]

In the bulk, the eigenmodes of induced polarization correspond to the upper and lower polariton (UP or LP), and the corresponding exciton state has a definite wave vector. In a slab, the translational symmetry is relaxed due to the finite thickness, but it tends to be recovered as we consider thicker slabs. Slabs are a good testing system to see how appropriately the development of wave vector conservation is described in a theoretical framework. In a thick bulk-like slab, a pump light with a given frequency will connect the biexciton states with polariton (UP or LP) states as shown in Fig. 5.6, where the propagation of the pump light is assumed to be in both forward and backward directions, corresponding to the internal reflection in the slab.

For fixed ω_{pu}, transitions to the biexciton states via UP and LP occur at different ω_{pr}, because of the finite biexciton mass. The pump–probe spectrum will show large signals when the intermediate step corresponds to one of these polariton states. Such pump-induced transitions obey the k-selection rule in the bulk, i.e.,

$$E_{\text{b}}(k_{\text{pr}} \pm k_{\text{pu}}) - \hbar\omega_{\text{pr}}(k_{\text{pr}}) = \hbar\omega_{\text{pu}}(\pm k_{\text{pu}}) , \tag{5.12}$$

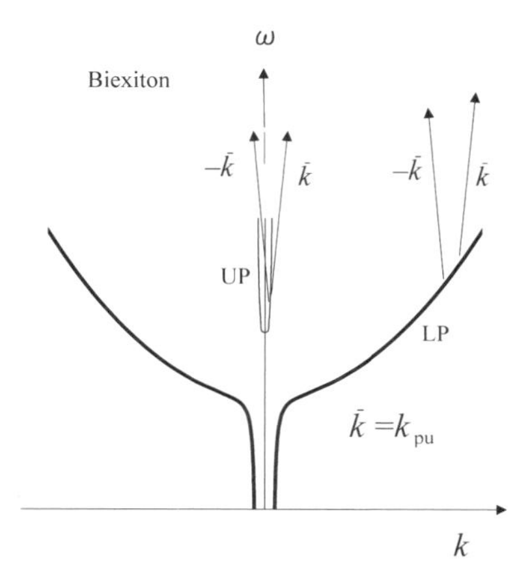

Fig. 5.6. Possible transitions between polariton and biexciton branches in the bulk for a pump light of a given frequency propagating in both forward and backward directions

where $E_b(k)$ represents the dispersion relation of the biexciton in the bulk, and both ω_{pr} and ω_{pu} lie on the exciton–polariton dispersion curve. For thin slabs, this type of sharp k-selection rule cannot be expected, as seen from Fig. 5.5.

For a slab with $d = 1$ μm, however, the large signal amplitude is seen to be concentrated at the frequencies expected from the bulk behavior, as shown in Fig. 5.7. Although the interference fringes indicate the remaining effect of finite thickness, the envelope has strong amplitude at the frequencies expected from the bulk behavior. Because of the condition (B) mentioned in the introduction, ω_{pr} can be on either the UP or the LP branch, while ω_{pu} is on the lower frequency region of the LP branch. Due to the presence of the forward and backward propagating pump light, there are two values of ω_{pr} satisfying the above equation for each of the processes via UP and LP. However, the solutions for the process via UP are too close to be spectrally distinguished. Therefore, there are only three main structures in Fig. 5.7. Together with the results for intermediate thicknesses, Figs. 5.5 and 5.7 clearly show the development of a bulk-like k-selection rule with increasing slab thickness.

Figure 5.8 shows the pump-induced change in the absorption spectrum as a function of ω_{pr} for $\hbar\omega_{pu}$ corresponding to the transition energy between a_4 and m_5 when $d = 200$ Å. There are three different features in this figure, i.e., two-photon absorption to a biexciton state, pump-induced broadening of an exciton level, and the optical Stark effect. They have different spectral shapes such as a simple peak (two-photon absorption), a first-derivative type due to the pump-induced shift of the exciton levels (i.e., the Stark effect) and a second-derivative form (broadening). The Stark shift of a size-quantized exciton level has a definite direction according to its position relative to the spectral region where the second-derivative form dominates. On the higher

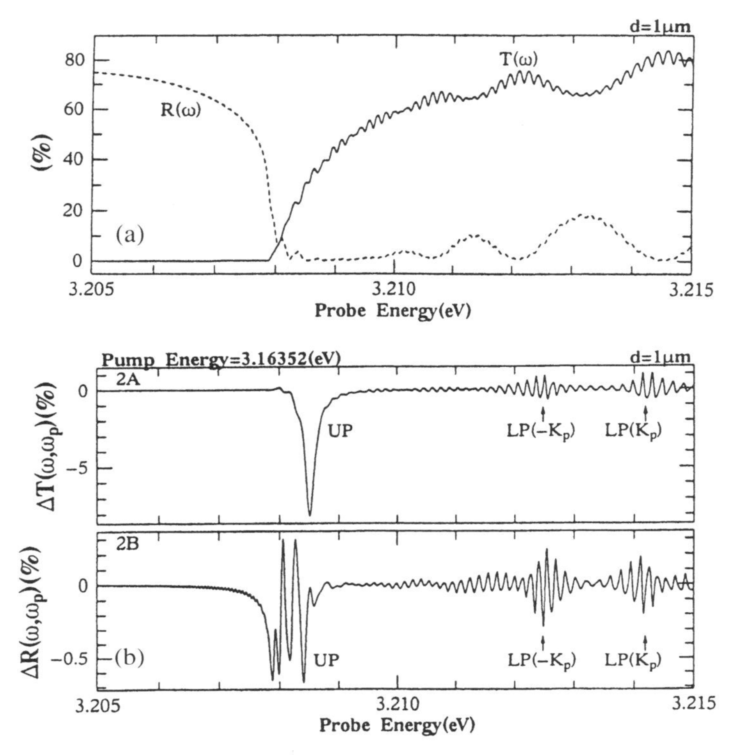

Fig. 5.7. (**a**) Reflectance and transmittance spectra of a $d = 1.0$ μm slab and (**b**) their change due to pump light with $\hbar\omega_{\rm pu} = 3.16352$ eV. Frequencies of resonances expected for the bulk corresponding to the transitions shown in Fig. 5.6 are indicated by LP(K) and UP [126]

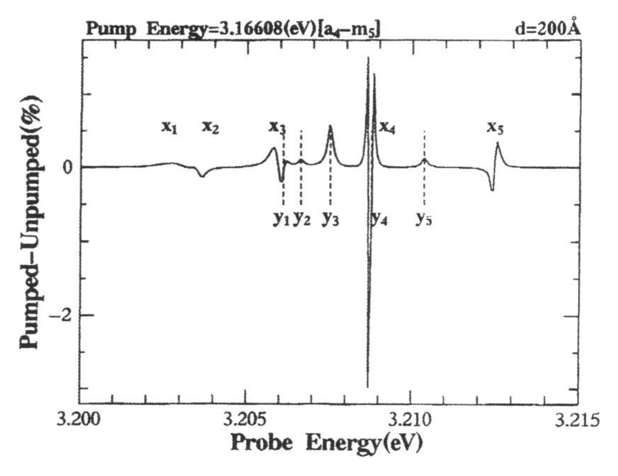

Fig. 5.8. Change in the absorption spectrum of Fig. 5.4 due to pump light with $\hbar\omega_{\rm pu} = 3.16608$ eV. *Vertical dashed lines* indicate the energies of two-photon absorption to the quantized biexciton levels [126]

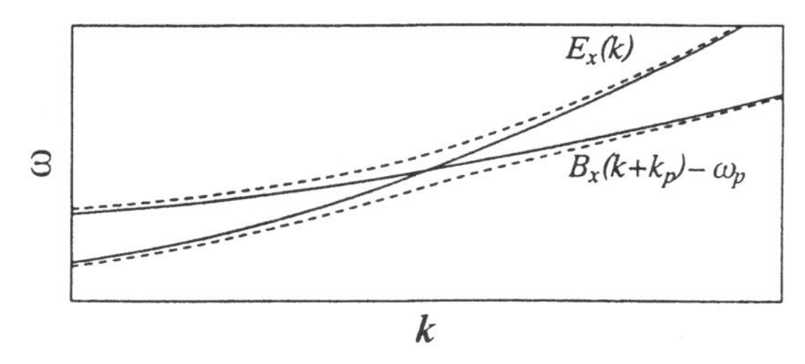

Fig. 5.9. Effective polariton branches (*dashed curves*) in the presence of pump light [126]

(lower) energy side of this region, the shift is towards higher (lower) energies. This can be understood according to the picture of a new excitation branch for the probe light in the presence of strong pump light. Hence, the probe light can produce matter excitations in two ways. These are the creation of excitons by the probe light alone and the creation of biexcitons by means of the pump light. The latter is in a sense equivalent to the emergence of a new excitation branch for the probe light, with dispersion

$$\hbar \Omega_{\mathrm{new}}(k) = E_{\mathrm{b}}(k + k_{\mathrm{pu}}) - \hbar\omega_{\mathrm{pu}}(k_{\mathrm{pu}}) \ . \tag{5.13}$$

Since this branch has an interaction with the exciton branch $E_{\mathrm{x}}(k)$ via the pump light, they form the mixed branch structure sketched in Fig. 5.9 (dashed lines).

In view of the fact that the exciton component is mainly seen by the probe light, the lower half of the lower branch and the upper half of the higher branch are expected to contribute to the pump-induced spectrum. Thus the pump-induced change in the spectrum shows different energy shifts in the size-quantized levels depending on their positions on these mixed branches.

5.2 Degenerate Four-Wave Mixing

In the nonlinear process of four-wave mixing (FWM), two incident waves with frequency and wave vector $(\omega_1, \boldsymbol{k}_1)$ and $(\omega_2, \boldsymbol{k}_2)$ produce a signal wave with $(2\omega_1 - \omega_2, 2\boldsymbol{k}_1 - \boldsymbol{k}_2)$. Since the propagation direction of the signal wave differs from those in the linear process, a signal observed in this direction is solely due to the nonlinear process. This is not the case in the pump–probe processes mentioned in the previous section, where the effect of the pump light is studied on the signal light, which also exists in the absence of pumping.

A special case of FWM is degenerate FWM (DFWM), where we have $\omega_1 = \omega_2$. In this case the signal light also has the same frequency, but its wave vector allows us to distinguish it from the incident waves. Because of the coincidence of all the frequencies of the incident and signal waves, the NIDORES effect in this nonlinear process can have multiply-resonant behavior, which is more resonant than the similar effect in the pump–probe process mentioned previously.

The third-order induced current density of DFWM is proportional to the product of two components of the pump field and the complex conjugate amplitude of the test field. In terms of the general expression for the third-order induced current density $\tilde{\boldsymbol{j}}^{(3)}$ in (2.121), the corresponding choice is that two of $(\omega_1, \omega_2, \omega_3)$ are the pump wave frequency $\omega_{\rm pu}$ and the other is the test wave frequency $-\omega_{\rm ts}$. This choice leads to the signal frequency

$$\omega_{\rm sig} = 2\omega_{\rm pu} - \omega_{\rm ts} \, . \tag{5.14}$$

If the sample is a slab in the xy plane, the lateral wave vector of the signal is given as

$$\boldsymbol{k}_{\|,\rm sig} = 2\boldsymbol{k}_{\|,\rm pu} - \boldsymbol{k}_{\|,\rm ts} \, . \tag{5.15}$$

For normally incident pump light, i.e., $\boldsymbol{k}_{\|,\rm pu} = 0$, we have $\boldsymbol{k}_{\|,\rm sig} = -\boldsymbol{k}_{\|,\rm pr}$. Hence the signal light is the conjugate wave of the test light.

The main point in the following discussion concerns the way we choose the appropriate condition for enhancing the DFWM signal by adjusting frequencies and sample size [123, 127]. The basic idea behind the NIDORES effect is that the two conditions,

- minimize the energy denominators,
- enhance the factors $\{\tilde{F}_{\mu\nu}(\omega)\}$ in the numerators of $\tilde{\boldsymbol{j}}^{(3)}$,

can be optimized cooperatively in order to obtain a well-enhanced signal. The first condition means choosing the terms of $\tilde{\boldsymbol{j}}^{(3)}$ with energy denominators that can be nearly zero within the freedom of the model system. Such terms can be chosen as the most (i.e., triply) resonant ones from the general expression (2.121). The states $|\mu\rangle$ and $|\sigma\rangle$ may be chosen as the one-exciton states $|{\rm x}n\rangle$ and $|{\rm x}m\rangle$, respectively, and $|\nu\rangle$ is either the ground state $|0\rangle$ or the biexciton state $|{\rm b}j\rangle$.

If $|\nu\rangle = |0\rangle$, the triply resonant terms are obtained from the four terms in (2.121), i.e., from the first one by the choice $\omega_1 = \omega_3 = -\omega_2 = \omega$ (i.e., $\Omega_3 = \omega + {\rm i}0^+$, $\Omega_2 = 2{\rm i}0^+$, $\Omega_1 = \omega + 3{\rm i}0^+$) as

$$\frac{\langle 0|\boldsymbol{I}|n\rangle \tilde{F}_{n0}(\omega)\tilde{F}_{m0}(\omega)\tilde{F}_{m0}(\omega)^*}{(\omega_{m0} - \Omega_3)(\omega_{00} - \Omega_2)(\omega_{n0} - \Omega_1)} \, , \tag{5.16}$$

from the fourth one by the choice $\omega_1 = \omega_3 = -\omega_2 = \omega$ as

$$\frac{\langle 0|\boldsymbol{I}|m\rangle \tilde{F}_{n0}(\omega)\tilde{F}_{m0}(\omega)\tilde{F}_{n0}(\omega)^*}{(\omega_{m0} - \Omega_3)(\omega_{mn} - \Omega_2)(\omega_{m0} - \Omega_1)} \, , \tag{5.17}$$

from the sixth one by the choice $\omega_1 = \omega_2 = -\omega_3 = \omega$ (i.e., $\Omega_3 = -\omega + {\rm i}0^+, \Omega_2 = 2{\rm i}0^+, \Omega_1 = \omega + 3{\rm i}0^+$) as

$$\frac{\langle 0|\boldsymbol{I}|m\rangle \tilde{F}_{n0}(\omega)\tilde{F}_{m0}(\omega)\tilde{F}_{n0}(\omega)^*}{(\omega_{0n}-\Omega_3)(\omega_{mn}-\Omega_2)(\omega_{m0}-\Omega_1)}, \tag{5.18}$$

and from the seventh term of (2.121) by the choice $\omega_1 = \omega_2 = -\omega_3 = \omega$ as

$$\frac{\langle 0|\boldsymbol{I}|m\rangle \tilde{F}_{m0}(\omega)\tilde{F}_{n0}(\omega)\tilde{F}_{n0}(\omega)^*}{(\omega_{0n}-\Omega_3)(\omega_{00}-\Omega_2)(\omega_{m0}-\Omega_1)}. \tag{5.19}$$

Considering the phenomenological damping constants γ and Γ for the diagonal and non-diagonal elements of the density matrix, respectively, and taking an additional restriction $n = m$ to make the resonance stronger, we can sum up the above four terms to give

$$-\frac{4\Gamma}{\gamma}\frac{1}{(\omega_{\mathrm{x}n}-\omega-\mathrm{i}\Gamma)[(\omega_{\mathrm{x}n}-\omega)^2+\Gamma^2]}. \tag{5.20}$$

For $|\nu\rangle = |\mathrm{b}j\rangle$, one of the biexciton states, there are four triply resonant terms from (2.121): the first term with the choice $-\omega_1 = \omega_2 = \omega_3 = \omega$ (i.e., $\Omega_3 = \omega + \mathrm{i}0^+$, $\Omega_2 = 2\omega + 2\mathrm{i}0^+$, $\Omega_1 = \omega + 3\mathrm{i}0^+$),

$$\frac{\langle 0|\boldsymbol{I}(\boldsymbol{r})|n\rangle \tilde{F}_{jn}(\omega)^*\tilde{F}_{jm}(\omega)\tilde{F}_{m0}(\omega)}{(\omega_{m0}-\Omega_3)(\omega_{j0}-\Omega_2)(\omega_{n0}-\Omega_1)}, \tag{5.21}$$

the second term with the choice $-\omega_1 = \omega_2 = \omega_3 = \omega$,

$$-\frac{\langle n|\boldsymbol{I}(\boldsymbol{r})|j\rangle \tilde{F}_{n0}(\omega)^*\tilde{F}_{jm}(\omega)\ \tilde{F}_{m0}(\omega)}{(\omega_{m0}-\Omega_3)(\omega_{j0}-\Omega_2)(\omega_{jn}-\Omega_1)}, \tag{5.22}$$

the third term with the choice $\omega_1 = -\omega_2 = \omega_3 = \omega$ (i.e., $\Omega_3 = \omega + \mathrm{i}0^+, \Omega_2 = 2\mathrm{i}0^+, \Omega_1 = \omega + 3\mathrm{i}0^+$),

$$-\frac{\langle n|\boldsymbol{I}(\boldsymbol{r})|j\rangle \tilde{F}_{n0}(\omega)^*\tilde{F}_{jm}(\omega)\tilde{F}_{m0}(\omega)}{(\omega_{m0}-\Omega_3)(\omega_{mn}-\Omega_2)(\omega_{jn}-\Omega_1)}, \tag{5.23}$$

and the fifth term with the choice $\omega_1 = \omega_2 = -\omega_3 = \omega$ (i.e., $\Omega_3 = -\omega + \mathrm{i}0^+, \Omega_2 = 2\mathrm{i}0^+, \Omega_1 = \omega + 3\mathrm{i}0^+$),

$$-\frac{\langle n|\boldsymbol{I}(\boldsymbol{r})|j\rangle \tilde{F}_{jm}(\omega)\tilde{F}_{m0}(\omega)\ \tilde{F}_{n0}(\omega)^*}{(\omega_{m0}-\Omega_3)(\omega_{mn}-\Omega_2)(\omega_{jn}-\Omega_1)}. \tag{5.24}$$

If we sum up these terms with the introduction of damping constants Γ and γ and assume $n = m$, the main contribution has energy denominator

$$\frac{1}{(\omega_{jn}-\omega-\mathrm{i}\Gamma)[(\omega_{\mathrm{x}n}-\omega)^2+\Gamma^2]}. \tag{5.25}$$

The second condition for the NIDORES effect mentioned above may be considered in a simplified way by using the results determined for the linear response to the pump light. The internal field should be determined from the third-order version of $\mathbf{S}\tilde{\boldsymbol{F}} = \tilde{\boldsymbol{F}}^{(0)}$, as discussed in Sect. 2.6, where the elements of the coefficient matrix $\mathbf{S}$ contain $\{\tilde{F}\}$ up to second order, while they are independent of $\{\tilde{F}\}$ in the case of linear response. If we are concerned with the effect of the pump light on the resonant structure of the linear response, the solutions to the third-order polynomial equations may be regarded as modified linear response solutions. The modification will be smaller for weaker pump light. In this sense, it would not be a bad approximation to replace the factors $\{\tilde{F}(\omega_{pu})\}$ in $\tilde{\boldsymbol{j}}^{(3)}$ by those for the linear response.

Taking the same model for numerical study as for the pump–probe effect discussed in the previous section Sect. 5.1.1, we can demonstrate the remarkable enhancement effect of DFWM, as expected from the NIDORES scheme.

Figure 5.10 shows the signal intensity as a function of beam energy $\hbar\omega$ and the thickness of the slab measured in terms of the number of lattice planes. The material parameters are for the Z_3 exciton in CuCl as before. Two sets of damping constants Γ and γ are used. This leads to a large difference

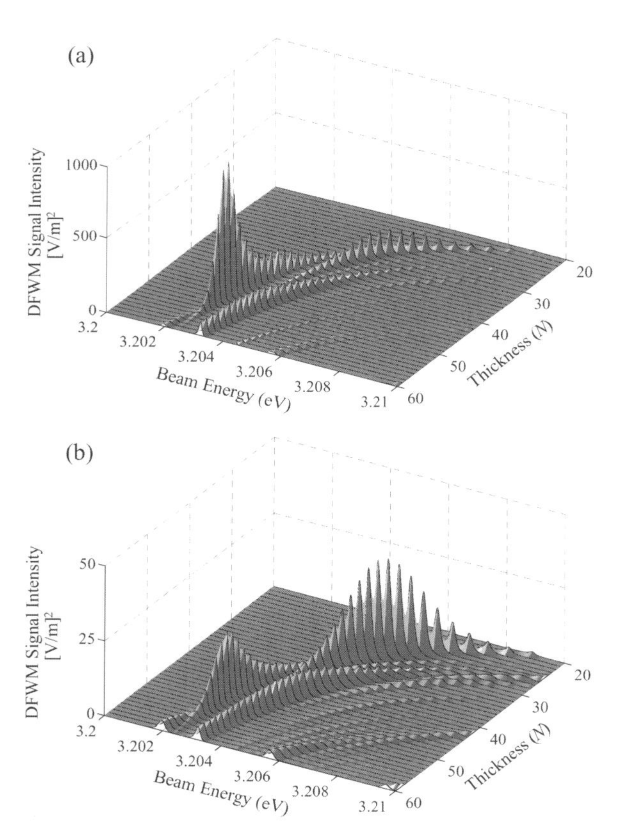

Fig. 5.10. DFWM Spectra for various thicknesses for two sets of non-radiative damping parameters (γ, Γ) in [meV]: (**a**) (0.02, 0.06) and (**b**) (0.04, 0.12). The amplitudes of the pump and test beams are 2.4×10^5 V/cm and 1.2×10^3 V/cm, respectively. Note the large difference in the ordinate units [123]

in the signal intensity. The smaller ones, which are realistic, as mentioned in Sect. 5.1.1, show a quite remarkable resonant enhancement. The largest enhancement is obtained for the transition from the ground to the second exciton level E_{x2}, which produces a large internal field at $N \approx 50$ (i.e., a thickness of about 300 Å). This is in accordance with the result of Fig. 4.10. It should be stressed again that this internal field has a spatial structure in the normal direction to the slab, which can efficiently induce non-electric-dipole type transitions to the level E_{x2}.

The NIDORES effect in DFWM is more effective than in pump–probe processes, because it is not only the test and pump beams but also the signal beam that is enhanced, and this in turn happens because all the beams have the same frequency.

Akiyama et al. have recently demonstrated this effect experimentally in a slab of GaAs of thickness about 100 nm [128, 129]. They used good quality samples of various thicknesses to see the double resonance feature with respect to size and beam energy. The result is shown in Fig. 5.11, where the double resonance feature is clearly visible. Figure 5.12 shows the comparison between theory and measurement for the maximum signal intensity versus sample thickness [130]. The good agreement can be taken as experimental proof of the NIDORES effect, which is one of the essential consequences of the nonlocal response theory.

The fact that the measurement is interpreted nicely by the NIDORES effect means that radiative damping is more important than non-radiative damping for the (weakly confined) size-quantized exciton in the GaAs slab. This leads us to expect a very rapid decay of the exciton states, even if they are resonantly excited. This has turned out to be the case, as demonstrated by the time-resolved DFWM measurement by Akiyama et al. [131]. Theoretical analysis of this temporal behavior has also been made [132] by means of the time-domain nonlocal response theory [133], which further supports the

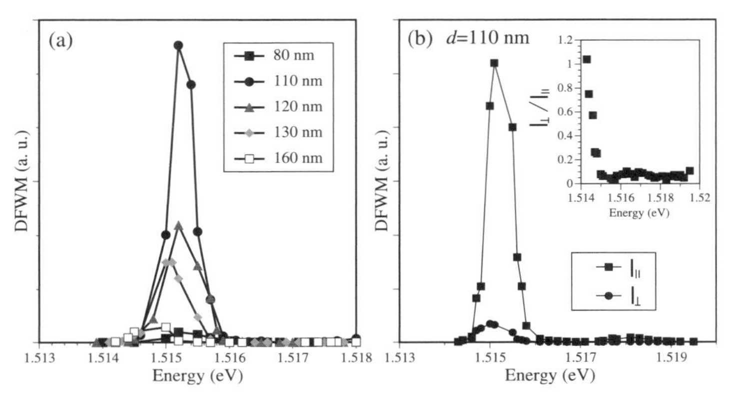

Fig. 5.11a,b. Double resonance feature in DFWM signals for GaAs slabs [130]

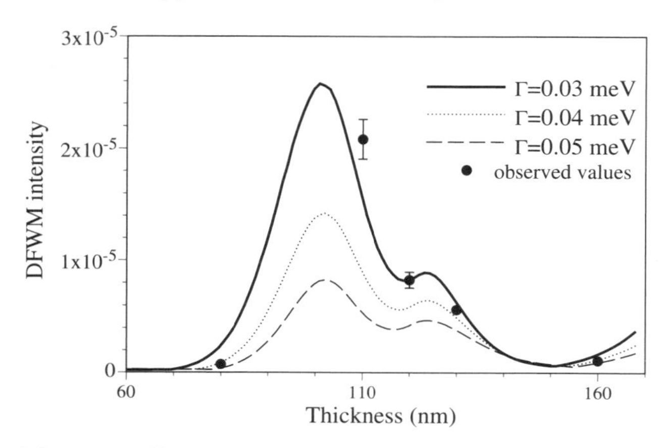

Fig. 5.12. Comparison between theory and experiment for DFWM in GaAs slabs. The DFWM intensity is the plot of the spectral maximum for each slab thickness [130]

dominance of radiative decay. This result will open up possiblities for applying the NIDORES effect to high-speed nonlinear devices.

5.3 Optical Bistability

Optical bistability is a nonlinear process where the output intensity is a two-valued function of the incident light intensity, and the signal shows an abrupt jump from one to the other as one sweeps the input power. The value of the input power at the point of the abrupt jump differs for the increasing or decreasing sweep process (hysteresis effect).

The study of optical bistability was quite popular during the 1980s [134]. The nonlocal point of view was not an important issue in such studies. Resonant conditions are not necessarily required, although they would lead to certain unexpected results. The discussion of resonant effects, especially in nanostructures, would require a nonlocal description. The nonlocal formalism developed here is general enough to describe resonant optical bistability. The nonlocal aspect has characteristic dependence on the spatial structure and amplitude of resonantly induced polarization, which is missing in the discussion of the bistability of local nonlinear media. We therefore think it worth providing an example of optical bistability from this new point of view.

The difference between local and nonlocal descriptions lies in the spatial structure of the internal field. In the nonlocal scheme, the internal field is the superposition of those due to all the resonant states, each of which has its characteristic structure. In the local case, on the other hand, the spatial structure is determined by the macroscopic susceptibilities, which gives the wavelength of the light of a given frequency. This difference in the character of the spatial structure would lead to a different feedback mechanism in the light amplitudes, which is the origin of bistability in general. When the internal field has strong spatial structure in a resonant regime, as mentioned throughout this chapter, the total internal field consisting of the sum of several components is significantly affected by the uneven feedback effect of the different components.

There is also a certain technically interesting feature in deriving this effect within the nonlocal formalism. Indeed, a straightforward numerical solution of the simultaneous third-order polynomial equations for different input intensities leads to the appearance of an abrupt change in the intensity of the output light signal. There are many sets of solutions to the simultaneous third-order polynomial equations. However, the numerical solutions quasi-continuously sought from the linear regime for very weak input power to the nonlinear regime allow us to find the right answer, connecting correctly to that of the linear regime. Hence, from a system of equations potentially containing many sets of solutions, we can obtain bistable behavior via numerical calculation without resorting to sophisticated techniques.

As a model system for numerical calculation, a model similar to the one in Sect. 5.1.1 was chosen, i.e., a slab consisting of bundles of 1D chains in the normal direction to the slab. Excitons can move within a chain via the transfer energy b, but the interchain transfer is assumed to be zero. The exciton–exciton interaction is neglected, except for the restriction from the Pauli principle, i.e., two excitons are forbidden on a single site.

As in the previous treatments in this chapter, we select most (triply) resonant terms in the induced third-order current density $\tilde{\boldsymbol{j}}^{(3)}$. There are two groups of such terms, i.e., those in (5.3) and (5.4) with the replacement $\omega_{\mathrm{pu}} = \omega_{\mathrm{pr}} = \omega$, and those arising from the first and the seventh terms of (2.122), which lead to [135]

$$\frac{4\Gamma|\tilde{F}_{\mathrm{n}0}(\omega)|^2\tilde{F}_{m0}(\omega)}{(\omega + \mathrm{i}\Gamma - E_{\mathrm{n}0})\gamma[\Gamma^2 + (E_{\mathrm{m}0} - \omega)^2]} \, . \tag{5.26}$$

In the present model, $\{|\mathrm{b}j\rangle\}$ in (5.3) and (5.4) correspond to unbound two-exciton states in a chain consisting of the sample slab. Using these terms as the third-order induced current density, we set up the third-order polynomial equations for $\{\tilde{F}\}$, and solve them numerically.

From the results of Sects. 5.1.1 and 5.2, $\tilde{F}_{n=2,0}(\omega)$ is known to be resonantly enhanced for a chain length d of about 30 nm. Thus in the frequency range of the resonant structure at $\hbar\omega = E_{n0} + \tilde{A}_{n0;0n}$ $(n = 2)$, we can expect a large nonlinear effect due to the enhanced $\tilde{F}_{n=2,0}(\omega)$ factor appearing in (5.3) and (5.4).

Figure 5.13 shows the linear transmittance spectrum of a slab of thickness $d = 227$ Å. The lowest peak corresponds to the resonant structure due to the enhancement of $\tilde{F}_{n=2,0}(\omega)$. As we increase the power of the light beam, the transmittance spectrum shows a frequency-dependent change. If we con-

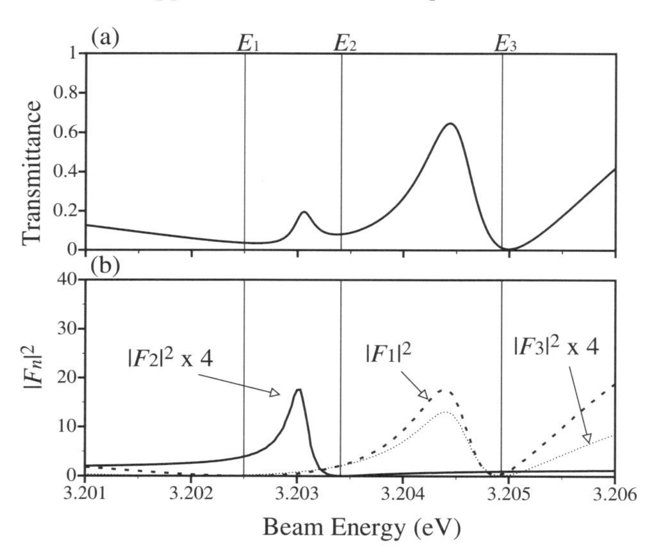

Fig. 5.13a,b. Linear transmittance spectrum of the slab with $d = 22.7$ nm. The E_i s are the size-quantized exciton energies [135]

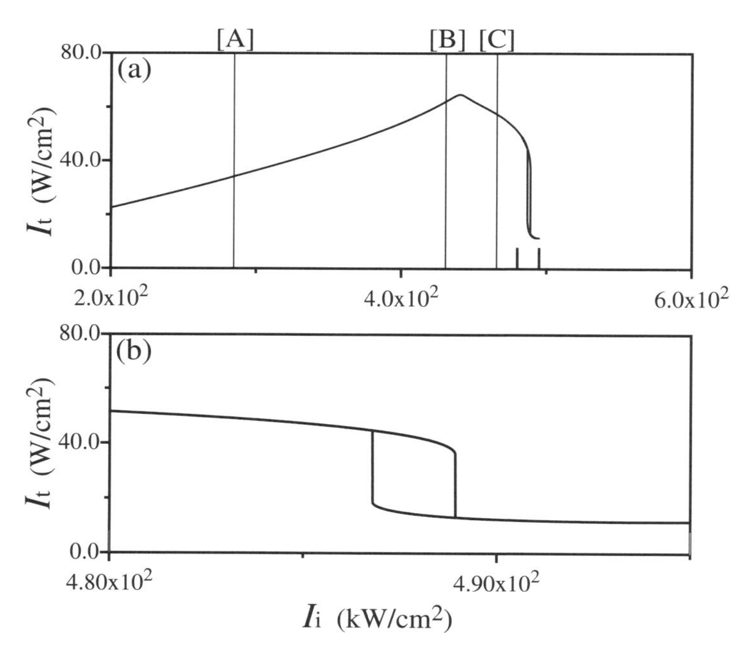

Fig. 5.14. Intensity of transmitted light as a function of incident intensity. (**b**) is the enlargement of (**a**) in the bistability region [135]

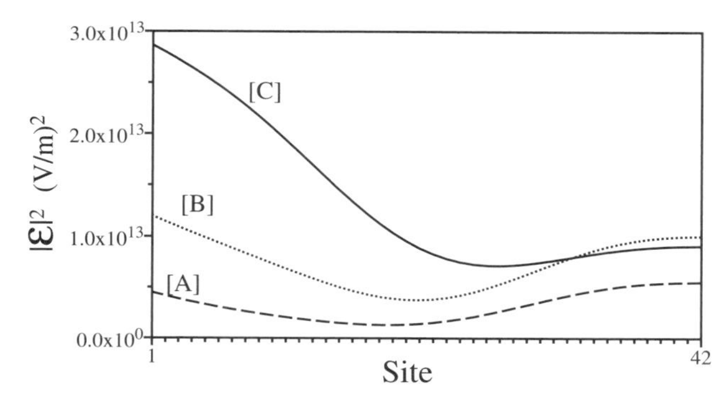

Fig. 5.15. Internal field patterns for the three incident intensities indicated in Fig. 5.14 [135]

centrate on the higher energy tail of the lowest energy peak in Fig. 5.13, its transmittance change shows a bistable behavior as in Fig. 5.14. The corresponding structure of the internal field is shown in Fig. 5.15, where the three curves correspond to the power values indicated by [A], [B], and [C] in Fig. 5.14. The last figure demonstrates the different power-dependent contributions of the components of the induced current density, as mentioned in the first part of this section.

It is remarkable that, if we deliberately neglect the nonlinear components due to (5.3) and (5.4), the bistable behavior still appears in the same way. This is a little surprising, because the nonlinear current density in this treatment is due solely to the ground and one-exciton states, without any contribution from the two-exciton states. This means that no induced absorption process is included, yet we nevertheless obtain the 'inverted form of bistability', i.e., less transmitted intensity for larger incident intensity. This is a peculiarity of the nonlocal nature of the nonlinear process under consideration.

References

Chapter 1

1. J.H. van Vleck: *Theory of Electric and Magnetic Susceptibilities* (Oxford University Press, 1932) p. 1
2. K. Cho: Prog. Theor. Phys. Suppl. **106** (1991) 225
3. M. Born and K. Huang: *Dynamical Theory of Crystal Lattices* (Clarendon Press, 1954) Sect. 19
4. H.A. Lorentz: *The Theory of Electrons* (Leipzig, 1916)
5. S.I. Pekar: Zh. Eksp. Teor. Fiz. **33** (1957) 1022 [Sov. Phys. JETP **6** (1957) 785]
6. J.L. Birman: *Excitons*, ed. by E.I. Rashba and M.D. Sturge (North Holland, 1982) p. 72
7. P. Halevi: *Spatial Dispersion in Solids and Plasmas*, ed. by P. Halevi (Elsevier, 1992) p. 339
8. V.M. Agranovich and V.L. Ginzburg: *Crystal Optics and Spatial Dispersion, and Excitons* (Springer, 1984)
9. F. Forstmann and R.R. Gerhardts: *Metal Optics Near the Plasma Frequency* (Springer, 1986)
10. C. Cohen-Tannoudji, J. Dupont-Roc, and G. Grynberg: *Photons and Atoms: Introduction to Quantum Electrodynamics* (Wiley, 1989) Chap. 1
11. Y. Toyozawa: *The Physics of Elementary Excitations*, ed. by S. Nakajima, Y. Toyozawa, and R. Abe (Springer, 1980) Chap. 2
12. K. Ehara and K. Cho: Solid State Commun. **44** (1982) 453
13. K. Cho: Proc. 14th Int. Conf. on *Physics of Semiconductors*, Edinburgh 1978, ed. by B.L.H. Wilson (Institute of Physics, Bristol and London) p. 841
14. K. Cho: J. Phys. Soc. Jpn. **68** (1999) 683
15. O. Keller: Phys. Rep. **268** (1996) 85
16. A. Stahl and I. Balslev: *Electrodynamics of the Semiconductor Band Edge*, Springer Tract in Mod. Phys. **110** (Springer, 1987)
17. See, for example, G. Czajkowski, M. Dressler, and F. Bassani: Phys. Rev. B **55** (1997) 5243; M. Dressler, F. Bassani, and G. Czajkowski: Eur. Phys. J. B **10** (1999) 681; L. Silvestri, G. Czajkowski, and F. Bassani: phys. stat. sol. (a) **175** (1999) 383

Chapter 2

18. W. Heitler: *The Quantum Theory of Radiation*, 3rd edn. (Clarendon Press, Oxford, 1954)
19. L.D. Landau and E.M. Lifshitz: *Quantum Mechanics: Nonrelativistic Theory*, 3rd edn. (Pergamon, 1977), Sect. 115
20. Y. Ohfuti and K. Cho: Phys. Rev. B **52** (1995) 4828; Errata: J. Lumin. **66/67** (1996) 94
21. S. Nakajima: Prog. Theor. Phys. **20** (1958) 948; R. Zwanzig: J. Chem. Phys. **33** (1960) 1338; H. Mori: Prog. Theor. Phys. **33** (1965) 423

Chapter 3

22. R. Fuchs, K.L. Kliewer, and W.J. Pardee: Phys. Rev. **150** (1966) 589
23. R.R. Chance, A. Prock, and R. Silbey: J. Chem. Phys. **60** (1974) 2744
24. S. Haroche and D. Kleppner: Phys. Today (Am. Inst. Phys.) **1** (1989) 24 and references therein
25. R.K. Chang and A.J. Campillo: *Optical Processes in Microcavities* (World Scientific, Singapore, 1996)
26. W.C. Chew: *Waves and Fields in Inhomogeneous Media* (Van Nostrand Reinhold, New York, 1990)
27. B.W. Batterman and H. Cole: Rev. Mod. Phys. **36** (1964) 681
28. U. van Bürck, R.L. Mössbauer, E. Gerdau, R. Rüffer, R. Hollatz, G.V. Smirnov, and J.P. Hannon: Phys. Rev. Lett. **59** (1987) 355
29. K. Cho: Phys. Rev. B **14** (1976) 4463
30. W.R. Heller and A. Marcus: Phys. Rev. **84** (1951) 809
31. Y. Onodera and Y. Toyozawa: J. Phys. Soc. Jpn. **22** (1967) 833
32. K. Cho: J. Phys. Soc. Jpn. **66** (1997) 2496
33. H. Ajiki and K. Cho: Proc. of 3rd Int. Conf on *Excitonic Processes in Condensed Matter*, Boston 1998, ed. by R.T. Willians and W.M. Yen (The Electrochemical Society, Pennington, NJ, 1999) pp. 262–237.
34. T. Takagahara: Phys. Rev. B **47** (1993) 4569
35. L.E. Brus: J. Chem. Phys. **80** (1984) 4403
36. Y. Kayanuma: Phys. Rev. B **38** (1988) 9797
37. H. Ajiki and K. Cho: Phys. Rev. B **62** (2000) 7402
38. R. Ruppin: J. Phys. Chem. Solids **50** (1989) 877
39. A.I. Ekimov, A.A. Onushchenko, M.E. Raikh, and A.L. Efros: Zh. Eksp. Teor. Fiz. **90** (1986) 1795 [Sov. Phys. JETP **63** (1986) 1054]
40. K. Cho and J. Ushida: Proc. Int. Symp. on *Elementary Processes in Excitations and Reactions on Solid Surfaces*, ed. by A. Okiji, H. Kasai, and K. Makoshi (Springer Series on Solid State Sciences 121) p. 193
41. O. Keller, M. Xiao, and S. Bozhevolnyi: Surf. Sci. **280** (1993) 217
42. J. Ushida: PhD Thesis (Osaka University, 1999)
43. J. Ushida, T. Ohta, and K. Cho: J. Phys. Soc. Jpn. **68** (1999) 2439
44. H. Haken: *Laser Theory* (Springer-Verlag, Heidelberg, 1984) p. 89
45. Y. Ohfuti, H. Fukutome, and K. Cho: J. Lumin. **72–74** (1997) 867
46. T. Minami, K. Sekine, H. Ajiki, and K. Cho: J. Lumin. **87–89** (2000) 378
47. R.J. Glauber: Phys. Rev Lett. **10** (1963) 84; Phys. Rev. **130** (1963) 2529; Phys. Rev. **131** (1963) 2766

48. G.S. Agarwal, G.S. Pattanayak, and E. Wolf: Phys. Rev. Lett. **27** (1971) 1022; Phys. Rev. B **8** (1973) 4768
49. R. Zeyher, J.L. Birman, and W. Brenig: Phys. Rev. B **6** (1972) 4613
50. A. Maradudin and D.L. Mills: Phys. Rev. B **7** (1973) 2787
51. A. D'Andrea and R. Del Sole: Phys. Rev. B **25** (1982) 3714
52. K. Cho and M. Kawata: J. Phys. Soc. Jpn. **54** (1985) 4431
53. K. Cho and H. Ishihara: J. Phys. Soc. Jpn. **59** (1990) 754
54. K. Cho: J. Phys. Soc. Jpn. **55** (1986) 4113
55. H. Ishihara and K. Cho: Phys. Rev. B **48** (1993) 7960.
56. J.J. Hopfield: Phys. Rev. **112** (1957) 1555
57. T. Takagahara and E. Hanamura: Phys. Rev. Lett. **56** (1986) 2533
58. E. Hanamura: Phys. Rev. **37** (1988) 1273
59. K. Cho: Proc. Int. Symp. on *Frontiers of Spintronics and Optics in Semiconductors*, Boston 2002, ed. by A.A. Efros, to be published
60. E.I. Rashba and G.E. Gurgenishvili: Sov. Phys. Solid State **4** (1962) 759
61. C.H. Henry and K. Nassau: Phys. Rev. B **1** (1970) 1628
62. E. Hanamura: Solid State Commun. **12** (1973) 951
63. T. Itoh, M. Furuyama, T. Ikehara, and C. Gourdon: Solid State Commun. **73** (1990) 271
64. Z.K. Tang, A. Yanase, T. Yasui, Y. Segawa, and K. Cho: Phys. Rev. Lett. **71** (1993) 1431
65. K. Cho, H. Ishihara, and T. Okada: Proc. Int. Meeting on *Optics of Excitons in Confined Systems*, ed. by A. D'Andrea, R. Del Sole, R. Girlanda, and A. Quattropani, Inst. Phys. Conf. Series **123**, p. 17
66. H. Ishihara and K. Cho: Phys. Rev. B **41** (1990) 1724
67. H. Ishihara and K. Cho: J. Nonlin. Opt. Phys. **1** (1992) 287
68. F.C. Spano and S. Mukamel: Phys. Rev. A**40** (1989) 5783
69. F.C. Spano and S. Mukamel: Phys. Rev. Lett. **66** (1991) 1197

Chapter 4

70. H. Ishihara and K. Cho: Solid State Commun. **89** (1994) 837
71. H. Ishihara and K. Cho: Phys. Rev. B **53** (1996) 15823
72. Y. Ohfuti and K. Cho: Phys. Rev. B **51** (1995) 14379
73. M. Orrit, C. Aslangul, and P. Kottis: Phys. Rev. B **25** (1982) 7263
74. J.J. Hopfield and D.G. Thomas: Phys. Rev. **132** (1963) 563
75. Z.K. Tang, A. Yanase, Y. Segawa, N. Matsuura, and K. Cho: Phys. Rev. B **52** (1995) 2640
76. C.F. Bohren and D.R. Huffman: *Absorption and Scattering of Light by Small Particles*, (John Wiley, New York, 1983)
77. K. Cho, M. Nishida, Y. Ohfuti, and L. Belleguie: J. Lumin. **60/61** (1994) 330
78. M. Nishida: MSc. Thesis (Osaka University, 1993, in Japanese)
79. H. Ajiki, T. Tsuji, K. Kawano, and K. Cho: Phys. Rev. B **66** (2002) 245322
80. E. Gerdau, R. Rüffer, H. Winkler, W. Tolksdorf, C.P. Klages, and J.P. Hannon: Phys. Rev. Lett. **54** (1985) 835; W. Sturhahn and E. Gerdau: Phys. Rev. B **49** (1994) 9285
81. G. Birkl, M. Gatzke, I.H. Deutsch, S.L. Rolston, and W.D. Phillips: Phys. Rev. Lett. **75** (1995) 2823; M. Weidemüller, A. Hemmerich, A. Görlitz, T. Esslinger, and T. Hänsch: Phys. Rev. Lett. **75** (1995) 4583
82. K. Cho and T. Arakawa: Mater. Sci. Eng. B **48** (1997) 71

83. T. Ikawa and K. Cho: J. Lumin. **87/89** (2000) 305; K. Cho and T. Ikawa: phys. stat. sol (b) **215** (1999) 281
84. T. Ikawa and K. Cho: J. Phys. Soc. Jpn. **71** (2002) 1381
85. T. Ikawa and K. Cho: Phys. Rev. B **66** (2002) 085338
86. E.L. Ivchenko, A.I. Nesvizhskii, and S. Jorda: Fiz. Tverd. Tela (St. Petersburg) **36** (1994) 2118 [Phys. Solid State **36** (1994) 1156]
87. See, for example, H. Haken: *Light* (North Holland, 1981) Vol. 1, Chap. 9
88. M. Naraschewski and A. Schenzle: Z. Physik D **33** (1995) 79
89. A. Ohishi: MSc. Thesis (Osaka University, 1996)
90. See, for example, A.J. Campillo, J.D. Eversole, and H-B. Lin: Phys. Rev. Lett. **67** (1991) 437
91. J. Ushida, T. Ohta, and K. Cho: J. Phys. Soc. Jpn. **68** (1999) 2439
92. M. Shinada and S. Sugano: J. Phys. Soc. Jpn. **21** (1966) 1936
93. Y. Masumoto, M. Matsuura, S. Tarucha, and H. Okamoto: Phys. Rev. B **32** (1985) 4275
94. M. Matsuura and Y. Shinozuka: J. Phys. Soc. Jpn. **53** (1984) 3138; G. Bastard, E.E. Mendez, L.L. Chang, and L. Esaki: Phys. Rev. B **26** (1982) 1974
95. N. Ochi, T. Shiotani, M. Yamanishi, Y. Honda, and I. Suemune: Appl. Phys. Lett. **58** (1991) 2735
96. Y. Yamamoto, S. Machida, K. Igeta, and G. Björk: *Coherence, Amplification and Quantum Effects in Semiconductor Lasers*, ed. by Y. Yamamoto (Wiley, New York, 1991) p. 561
97. C. Weisbuch, M. Nishioka, A. Ishikawa, and Y. Arakawa: Phys. Rev. Lett. **69** (1992) 3314
98. E.T. Jaynes and F.W. Cummings: Proc. IEEE **51** (1963) 89
99. G.S. Agarwal: J. Opt. Soc. Am. **2** (1985) 480
100. M. Kuwata-Gonokami, S. Inoue, H. Suzuura, M. Shirane, and R. Shimano: Phys. Rev. Lett. **79** (1997) 1341
101. P.G. Savviddis, J.J. Baumberg, R.M. Stevenson, M.S. Skolnick, D.M. Whittaker, and J.S. Roberts: Phys. Rev. Lett. **84** (2000) 1547; Phys. Rev. B **64** (2001) 75311
102. A.L. Ivanov and L.V. Keldysh: Zh. Eksp. Teor. Fiz. **84** (1983) 404 [Sov. Phys. JETP **57** (1983) 234]
103. A.L. Ivanov and P.B. Littlewood: Phys. Rev. Lett. **87** (2001) 136403
104. M. Nieto-Vesperinas and N. Garcia (Eds.): *Optics at the Nanometer Scale*, NATO ASI Series E319 (Kluwer Academic, Dordrecht, 1996)
105. S. Kawata, M. Ohtsu, and M. Irie (Eds.): *Nano-Optics* (Springer, 2002)
106. See, for example, T. Sugiura and T. Okada: Proc. SPIE, **3260** (1998) 4
107. F. Zenhausern, Y. Maartin, and H.K. Wickramashinge: Science **269** (1995) 1083
108. S. Kawata, K. Tanaka, and N. Takahashi: reference [105], Chap. 2
109. E. Betzig and J.K. Trautmann: Science **257** (1992) 189
110. U. Ch. Fischer and D.W. Pohl: Phys. Rev. Lett. **23** (1989) 458
111. K. Cho, Y. Ohfuti, and K. Arima: Jpn. J. Appl. Phys. **34**, Suppl. 34-1 (1995) 267
112. J. Ushida and K. Cho: Mol. Cryst. Liq. Cryst. **314** (1998) 215
113. J. Ushida: PhD. Thesis (Osaka University, 1999)
114. K. Cho, Y. Ohfuti, and K. Arima: Surf. Sci. **363** (1996) 378
115. H.S. Sözüer and J.W. Haus: Phys. Rev. B **45** (1992) 13962
116. K.M. Ho, C.T. Chan, and C.M. Soukoulis: Phys. Rev. Lett. **65** (1990) 3152
117. H.S. Sözüer and J.W. Haus: J. Opt. Soc. Am. **10** (1993) 296

118. K. Sakoda: *Optical Properties of Photonic Crystals*, Springer Series on Optical Sciences (Springer-Verlag, 2001); C.M. Soukoulis (Ed.): *Photonic Crystals and Light Localization in the 21st Century*, Nato Science Series C: Mathematical and Physical Science, Vol. 563 (Kluwer Academic, 2001)
119. K. Cho and T. Ikawa: phys. stat. sol. (a) **190** (2002) 401
120. K. Cho, T. Hirai, and T. Ikawa: J. Lumin. **100** (2002) 269

Chapter 5

121. N. Bloembergen: *Nonlinear Optics* (Benjamin, 1965)
122. N. Bloembergen, R.K. Chang, S.S. Jha, and C.H. Lee: Phys. Rev. **174** (1968) 813
123. H. Ishihara, T. Amakata, and K. Cho: Phys. Rev. B **65** (2002) 035305
124. Y.R. Shen: *The Principles of Nonlinear Optics* (John Wiley, New York, 1984) p. 13
125. T. Mita and N. Nagasawa: Solid State Commun. **44** (1982) 1003
126. N. Matsuura and K. Cho: J. Phys. Soc. Jpn. **64** (1995) 651; K. Cho and N. Matsura: Proc. Int. Conf. on Excitonic Processes in Condensed Matter, SPIE **2362** (1994) 151
127. H. Ishihara, T. Amakata, and K. Cho: J. Lumin. **87–89** (2000) 850
128. K. Akiyama, N. Tomita, Y. Nomura, and T. Isu: Appl. Phys. Lett. **75** (1999) 475
129. K. Akiyama, N. Tomita, Y. Nomura, T. Isu, H. Ishihara, and K. Cho: Physica E**7** (2000) 661
130. H. Ishihara, K. Cho, K. Akiyama, N. Tomita, Y. Nomura, and T. Isu: Phys. Rev. Lett. **89** (2002); Proc. 25th Int. Conf. Physics of Semiconductors (2001) p. 593
131. K. Akiyama, N. Tomita, T. Nishimura, Y. Nomura, and T. Isu: Abstracts 7th Int. Workshop on *Femtosecond Technology* (FST2000, Tsukuba) p. 69
132. H. Ishihara, K. Cho, K. Akiyama, N. Tomita, and T. Isu: phys. stat. sol. (a) **190** (2002) 849
133. H. Ishihara: Mater. Sci. Eng. B **48** (1997) 75; J. Lumin. **87–89** (2000) 905
134. See, for example, P. Mandel, S.D. Smith, and B.S. Wherrett (Eds.): *From Optical Bistability Towards Optical Computing* (North-Holland, 1987)
135. H. Ishihara: QELS Technical Digest Series **16** (1995) p. 2; Phys. Rev. B **67** (2003) in print

Index

Printing (Computer to Plate): Saladruck Berlin
Binding: Stürtz AG, Würzburg

Springer Series in Solid-State Sciences

Editors: M. Cardona P. Fulde K. von Klitzing H.-J. Queisser

Springer Series in Solid-State Sciences

47 **Statics and Dynamics of Nonlinear Systems**
Editors: G. Benedek, H. Bilz, and R. Zeyher

48 **Magnetic Phase Transitions**
Editors: M. Ausloos and R. J. Elliott

49 **Organic Molecular Aggregates**
Electronic Excitation and Interaction Processes
Editors: P. Reineker, H. Haken, and H. C. Wolf

50 **Multiple Diffraction of X-Rays in Crystals**
By Shih-Lin Chang

51 **Phonon Scattering in Condensed Matter**
Editors: W. Eisenmenger, K. Laßmann, and S. Döttinger

52 **Superconductivity in Magnetic and Exotic Materials** Editors: T. Matsubara and A. Kotani

53 **Two-Dimensional Systems, Heterostructures, and Superlattices**
Editors: G. Bauer, F. Kuchar, and H. Heinrich

54 **Magnetic Excitations and Fluctuations**
Editors: S. W. Lovesey, U. Balucani, F. Borsa, and V. Tognetti

55 **The Theory of Magnetism II** Thermodynamics and Statistical Mechanics By D. C. Mattis

56 **Spin Fluctuations in Itinerant Electron Magnetism** By T. Moriya

57 **Polycrystalline Semiconductors**
Physical Properties and Applications
Editor: G. Harbeke

58 **The Recursion Method and Its Applications**
Editors: D, G. Pettifor and D. L. Weaire

59 **Dynamical Processes and Ordering on Solid Surfaces** Editors: A. Yoshimori and M. Tsukada

60 **Excitonic Processes in Solids**
By M. Ueta, H. Kanzaki, K. Kobayashi, Y. Toyozawa, and E. Hanamura

61 **Localization, Interaction, and Transport Phenomena** Editors: B. Kramer, G. Bergmann, and Y. Bruynseraede

62 **Theory of Heavy Fermions and Valence Fluctuations** Editors: T. Kasuya and T. Saso

63 **Electronic Properties of Polymers and Related Compounds**
Editors: H. Kuzmany, M. Mehring, and S. Roth

64 **Symmetries in Physics** Group Theory Applied to Physical Problems 2nd Edition
By W. Ludwig and C. Falter

65 **Phonons: Theory and Experiments II**
Experiments and Interpretation of Experimental Results By P. Brüesch

66 **Phonons: Theory and Experiments III**
Phenomena Related to Phonons
By P. Brüesch

67 **Two-Dimensional Systems: Physics and New Devices**
Editors: G. Bauer, F. Kuchar, and H. Heinrich

68 **Phonon Scattering in Condensed Matter V**
Editors: A. C. Anderson and J. P. Wolfe

69 **Nonlinearity in Condensed Matter**
Editors: A. R. Bishop, D. K. Campbell, P. Kumar, and S. E. Trullinger

70 **From Hamiltonians to Phase Diagrams**
The Electronic and Statistical-Mechanical Theory of sp-Bonded Metals and Alloys By J. Hafner

71 **High Magnetic Fields in Semiconductor Physics**
Editor: G. Landwehr

72 **One-Dimensional Conductors**
By S. Kagoshima, H. Nagasawa, and T. Sambongi

73 **Quantum Solid-State Physics**
Editors: S. V. Vonsovsky and M. I. Katsnelson

74 **Quantum Monte Carlo Methods in Equilibrium and Nonequilibrium Systems** Editor: M. Suzuki

75 **Electronic Structure and Optical Properties of Semiconductors** 2nd Edition
By M. L. Cohen and J. R. Chelikowsky

76 **Electronic Properties of Conjugated Polymers**
Editors: H. Kuzmany, M. Mehring, and S. Roth

77 **Fermi Surface Effects**
Editors: J. Kondo and A. Yoshimori

78 **Group Theory and Its Applications in Physics**
2nd Edition
By T. Inui, Y. Tanabe, and Y. Onodera

79 **Elementary Excitations in Quantum Fluids**
Editors: K. Ohbayashi and M. Watabe

80 **Monte Carlo Simulation in Statistical Physics**
An Introduction 4th Edition
By K. Binder and D. W. Heermann

81 **Core-Level Spectroscopy in Condensed Systems**
Editors: J. Kanamori and A. Kotani

82 **Photoelectron Spectroscopy**
Principle and Applications 2nd Edition
By S. Hüfner

83 **Physics and Technology of Submicron Structures**
Editors: H. Heinrich, G. Bauer, and F. Kuchar

84 **Beyond the Crystalline State** An Emerging Perspective By G. Venkataraman, D. Sahoo, and V. Balakrishnan

85 **The Quantum Hall Effects**
Fractional and Integral 2nd Edition
By T. Chakraborty and P. Pietiläinen

86 **The Quantum Statistics of Dynamic Processes**
By E. Fick and G. Sauermann

87 **High Magnetic Fields in Semiconductor Physics II**
Transport and Optics Editor: G. Landwehr

88 **Organic Superconductors** 2nd Edition
By T. Ishiguro, K. Yamaji, and G. Saito

89 **Strong Correlation and Superconductivity**
Editors: H. Fukuyama, S. Maekawa, and A. P. Malozemoff

Springer Series in Solid-State Sciences

Editors: M. Cardona P. Fulde K. von Klitzing H.-J. Queisser